U.S.NRC

United States Nuclear Regulatory Commission

Protecting People and the Environment

NUREG-1556
Vol. 13, Rev. 1

I0484554

Consolidated Guidance About Materials Licenses

Program-Specific Guidance About Commercial Radiopharmacy Licenses

Final Report

Manuscript Completed: November 2007
Date Published: November 2007

Prepared by
D.E. White, J.F. Katanic, D.B. Howe

Office of Federal and State Materials and
Environmental Management Programs

ABSTRACT

On August 8, 2005, the Energy Policy Act of 2005 (EPAct) gave NRC new regulatory authority over additional byproduct material. This new byproduct material now also includes naturally occurring materials, such as discrete sources of Radium-226, and accelerator-produced radioactive materials (NARM). This revision of NUREG-1556, Vol. 13, adds guidance needed to license commercial radiopharmacies as a result of the regulatory changes made by the EPAct and the NARM rule, "Requirements for Expanded Definition of Byproduct Material."

This guidance document contains information that is intended to assist applicants for commercial radiopharmacy licenses in preparing their license applications. In particular, it describes the type of information needed to complete NRC Form 313, "Application for Materials License." This document both describes the methods acceptable to NRC license reviewers in implementing the regulations and the techniques used by the reviewers in evaluating the application to determine if the proposed activities are acceptable for licensing purposes.

Paperwork Reduction Act Statement

This NUREG contains information collection requirements that are subject to the Paperwork Reduction Act of 1995 (44 U.S.C. 3501 et seq.). These information collections were approved by the Office of Management and Budget, approval numbers 3150-0044; 3150-0014; 3150-0017; 3150-0016; 3150-0001; 3150-0010; 3150-0158; 3150-0020; 3150-0009; 3150-0008; 3150-0132; and 3150-0120.

Public Protection Notification

The NRC may not conduct or sponsor, and a person is not required to respond to, a request for information or an information collection requirement unless the requesting document displays a currently valid OMB control number.

CONTENTS

CONTENTS

APPENDICES

FIGURES

TABLES

FOREWORD

The United States Nuclear Regulatory Commission (NRC) is using Business Process Redesign (BPR) techniques to redesign its materials licensing process. This effort is described in NUREG-1539, "Methodology and Findings of NRC's Materials Licensing Process Redesign," dated April 1996. A critical element of the new process is consolidating and updating numerous guidance documents into a NUREG series of reports. Below is a list of volumes currently included in the NUREG-1556 series: ·

Vol. No.	Volume Title	Status
1, Rev. 1	Program-Specific Guidance About Portable Gauge Licenses	Final Report
2	Program-Specific Guidance About Radiography Licenses	Final Report
3, Rev. 1	Applications for Sealed Source and Device Evaluation and Registration	Final Report
4	Program-Specific Guidance About Fixed Gauge Licenses	Final Report
5	Program-Specific Guidance About Self-Shielded Irradiator Licenses	Final Report
6	Program-Specific Guidance About 10 CFR Part 36 Irradiator Licenses	Final Report
7	Program-Specific Guidance About Academic, Research and Development, and Other Licenses of Limited Scope	Final Report
8	Program-Specific Guidance About Exempt Distribution Licenses	Final Report
9, Rev.2	Program-Specific Guidance About Medical Use Licenses	Final Report
10	Program-Specific Guidance About Master Materials Licenses	Final Report
11	Program-Specific Guidance About Licenses of Broad Scope	Final Report
12	Program-Specific Guidance About Possession Licenses for Manufacturing and Distribution	Final Report
13, Rev. 1	Program-Specific Guidance About Commercial Radiopharmacy Licenses	Final Report
14	Program-Specific Guidance About Well Logging, Tracer, and Field Flood Study Licenses	Final Report
15	Guidance About Changes of Control and About Bankruptcy Involving Byproduct, Source, or Special Nuclear Materials Licenses	Final Report
16	Program-Specific Guidance About Licenses Authorizing Distribution to General Licensees	Final Report
17	Program-Specific Guidance About Special Nuclear Material of Less Than Critical Mass Licenses	Final Report
18	Program-Specific Guidance About Service Provider Licenses	Final Report

Vol. No.	Volume Title	Status
19	Guidance For Agreement State Licensees Proposing to Work in NRC Jurisdiction (Non-Agreement States, Areas of Exclusive Federal Jurisdiction, or Offshore Waters) and Guidance For NRC Licensees Proposing to Work in Agreement State Jurisdiction (Reciprocity)	Final Report
20	Guidance About Administrative Licensing Procedures	Final Report
21	Program-Specific Guidance About Possession Licenses for Production of Radioactive Materials Using an Accelerator	Final Report

The current document, NUREG-1556, Vol. 13, Rev. 1, "Consolidated Guidance About Materials Licenses: Program-Specific Guidance About Commercial Radiopharmacies," is intended for use by applicants, licensees, NRC license reviewers, and other NRC personnel. This revision updates the previous information contained in NUREG-1556, Vol. 13, to address changes to NRC regulations due to the Energy Policy Act of 2005 (EPAct). The original NUREG-1556, Vol. 13, combined and superseded the guidance for applicants and licensees previously found in "Draft Regulatory Guide DG-0006" (previously issued as FC 410-4), "Guide for the Preparation of Applications for Commercial Nuclear Pharmacy Licenses" (March 1997) and Draft Standard Review Plan 85-14, "Standard Review Plan for Applications for Nuclear Pharmacy Licenses." This report also contains pertinent information found in Technical Assistance Requests and Information Notices, as listed in Appendix B.

On August 8, 2005, the EPAct was signed into law, which required NRC to amend its regulations to include jurisdiction over certain radium sources, accelerator-produced radioactive materials, and certain naturally occurring radioactive material. The EPAct expanded the Atomic Energy Act of 1954 definition of byproduct material to include any discrete source of radium-226 (Ra-226), any material made radioactive by use of a particle accelerator, and any discrete source of naturally occurring radioactive material (other than source material) that the Commission determines will pose a threat to the public health and safety or the common defense and security. This expanded definition includes the material that is produced, extracted, or converted after extraction for use for a commercial, medical, or research activity. The guidance contained in NUREG-1556, Vol. 13, Rev. 1, includes updated guidance on requirements for licensing the accelerator-produced radioactive materials and discrete sources of Ra-226 now included in the expanded definition of byproduct material.

This report takes a risk-informed, performance-based approach to licensing commercial radiopharmacies. It identifies the information needed from an applicant seeking to possess and use byproduct materials during the preparation and distribution of radioactive drugs and in the distribution of radiochemicals, sealed sources, and *in vitro* test kits. It does not address the production of radionuclides by an accelerator. Information for the production of radionuclides is provided in NUREG-1556, Vol. 21, "Consolidated Guidance About Materials Licenses: Program-Specific Guidance About Possession License for Production of Radioactive Materials Using an Accelerator."

A team composed of NRC staff from headquarters and regional offices prepared this document, drawing on their collective experience in radiation safety in general and as specifically applied to commercial radiopharmacy. A representative of NRC's Office of the General Counsel provided a legal perspective.

This report represents a step in the transition from the current paper-based process to the new electronic process. This document is available on the Internet at the following address: http://www.nrc.gov/reading-rm/doc-collections/nuregs/staff/sr1556/v13/.

NUREG-1556, Vol. 13, Rev. 1, is not a substitute for NRC regulations, and compliance is not required. The approaches and methods described in this report are provided for information only. Methods and solutions different from those described in this report will be acceptable if they provide a basis for the staff to make the determination needed to issue or continue a license.

Charles L. Miller

Charles L. Miller, Office Director
Office of Federal and State Materials and
  Environmental Management Programs

ACKNOWLEDGMENTS

The writing team thanks the individuals listed below for assisting in the development and review of the report. All participants provided valuable insights, observations, and recommendations.

The team would like to thank NRC Regional Offices and all of the States that provided comments and technical information that assisted in the development of this report. The team also thanks Justine Cowan, Loleta Dixon, Agi Seaton, and Roxanne Summers of Computer Sciences Corporation.

The Participants for this Revision

Bailey, Edgar D.
Bakhsh, Sarah R.
Beardsley, Michelle R.
Howe, Donna-Beth
Katanic, Janine F.
Taylor, Torre M.
Tobin, Jennifer C.
White, Duane E.
Williamson, Michael K.

The Participants for the Original Version

Cameron, Jamnes L.
Camper, Larry W.
Cool, Donald A.
Henderson, Pamela J.
Hickey, John W.
Hosey, Charles M.
Howe, Donna-Beth
Howell, Linda L.
Kinneman, John D.
Merchant, Sally L.
Montgomery, James L.
Phillips, Monte P.
Piccone, Josephine M.
Roe, Mary Louise
Schwartz, Maria E.
Treby, Stuart A.

ABBREVIATIONS

ACMUI   Advisory Committee on the Medical Uses of Isotopes

ALARA   as low as is reasonably achievable

ALI   annual limit on intake

ANP   authorized nuclear pharmacist

ANSI   American National Standards Institute

AU   authorized user

bkg   background

BPR   Business Process Redesign

Bq   becquerel

CDE   committed dose equivalent

CEDE   committed effective dose equivalent

CFR   Code of Federal Regulations

Ci   curie

cm   centimeter

cpm   counts per minute

DAC   derived air concentration

DDE   deep-dose equivalent

DFP   decommissioning funding plan

DIS   decay in storage

DOE   United States Department of Energy

DOT   United States Department of Transportation

dpm   disintegrations per minute

dpm/cm^2 disintegrations per minute per square centimeter

DU   depleted uranium

EDE   effective dose equivalent

EPAct   Energy Policy Act of 2005

FA   financial assurance

FDA   United States Food and Drug Administration

FSME   Office of Federal and State Materials and Environmental Programs

ABBREVIATIONS

G-M   Geiger-Mueller

GBq   Giga becquerel

GL   generic letter

GPO   Government Printing Office

IN   Information Notice

IP   Inspection Procedure

LLEA   local law enforcement agencies

LSC   Liquid Scintillation Counter

MC   Manual Chapter

mCi   millicurie

mGy   milliGray

MDA   minimum detectable activity

MOU   Memorandum of Understanding

mR   milliroentgen

mrem   millirem

mrem/hr   millirem per hour

mSv   millisievert

mSv/hr   millisievert per hour

NaI   sodium iodide

NARM   Naturally Occurring and Accelerator-Produced Radioactive Material

NCRP   National Council on Radiation Protection and Measurements

NIST   National Institute of Standards and Technology

NMSS   Office of Nuclear Materials Safety and Safeguards

NRC   Nuclear Regulatory Commission

NVLAP   National Voluntary Laboratory Accreditation Program

OSL   optically stimulated luminescence

P&GD   Policy and Guidance Directive

PET   Positron Emission Tomography

QA   quality assurance

R   roentgen

RG   Regulatory Guide

RQ   reportable quantity

RSO   Radiation Safety Officer

SI   International System of Units (abbreviated SI from the French, Le Système Internationale d'Unités)

SRP   Standard Review Plan

SSDR   Sealed Source and Device Registry

std   standard

Sv   sievert

TAR   Technical Assistance Request

TEDE   total effective dose equivalent

TI   transportation index

TLD   thermoluminescent dosimeters

1   PURPOSE OF REPORT

This report provides guidance to an applicant applying for a commercial radiopharmacy license, as well as providing NRC with the appropriate criteria for evaluating such applications. Within this document, the terms "byproduct material," "licensed material," and "radioactive material" are used interchangeably. In addition, the phrases or terms, "commercial radiopharmacy," "radiopharmacy," and "nuclear pharmacy," are used interchangeably.

Commercial radiopharmacy licenses are those licenses issued by the NRC, pursuant to 10 CFR Part 30 and 10 CFR 32.72, for the possession and use of radioactive materials for the manufacture, preparation, or transfer for commercial distribution of radioactive drugs containing byproduct material for medical use under 10 CFR Part 35. Within this document, preparation includes the making of radiopharmaceuticals from reagent kits (e.g., technetium-99m MAA (macroaggregated albumin)), and from raw materials (e.g., the compounding of radioiodine capsules for diagnostic and therapeutic medical use or Positron Emission Tomography (PET) radiopharmaceuticals for medical use). Commercial radiopharmacies may also be authorized to transfer for commercial distribution *in vitro* test kits described in 10 CFR 31.11, radiopharmaceuticals to licensees authorized to possess them for other than human medical use (e.g., veterinary medicine and research licensees), and radiochemicals to those licensees authorized to possess them, pursuant to 10 CFR Part 30. In addition, 10 CFR Part 30 authorizes radiopharmacies to redistribute (transfer) sealed sources for calibration and medical use initially distributed by a manufacturer licensed pursuant to 10 CFR 32.74.

Specific guidance for applicants requesting the production of radioactive material using an accelerator (e.g., PET radiopharmacies) is included in NUREG-1556, Vol. 21, "Consolidated Guidance About Materials Licenses: Program-Specific Guidance About Possession Licenses for Production of Radioactive Materials Using an Accelerator." Note that this guidance (Vol. 13) should be used for the activities that take place after the radiochemical is produced, which would include compounding the radiochemical to a radiopharmaceutical.

Also, specific guidance for applicants requesting authorization to manufacture and initially distribute molybdenum-99/technetium-99m generators, *in vitro* kits, radiochemicals and sealed sources is included in NUREG-1556, Vol. 12, "Consolidated Guidance About Materials Licenses: Program-Specific Guidance About Possession Licenses for Manufacturing and Distribution," and is not within the scope of this guidance for commercial radiopharmacies. These activities require specific NRC or Agreement State authorization and must be included on a specific license.

Furthermore, specific guidance for applicants requesting authorization to manufacture, distribute, and redistribute radioactive drugs to persons exempt from licensing (e.g., carbon-14 tagged urea) is included in NUREG-1556, Vol. 8, "Consolidated Guidance About Materials Licenses: Program-Specific Guidance About Exempt Distribution Licenses," and also is not within the scope of this guidance. These activities require specific NRC authorization and require the issuance of a separate license for exempt distribution.

PURPOSE OF REPORT

This report identifies the information needed to complete NRC Form 313, "Application for Materials License" (See Appendix A), for the use of byproduct materials in commercial radiopharmacies. The information collection requirements in 10 CFR Part 30 and NRC Form 313 have been approved under the Office of Management and Budget (OMB) Clearance Nos. 3150-0017 and 3150-0120, respectively.

The format within this document for each item of technical information is as follows:

- **Regulations** — references the regulations applicable to the item;
- **Criteria** — outlines the criteria used to judge the adequacy of the applicant's response;
- **Discussion** — provides additional information on the topic sufficient to meet the needs of most readers; and
- **Response from Applicant** — provides suggested response(s), offers the option of an alternative reply, or indicates that no response is needed on that topic during the licensing process.

Notes and References are self-explanatory and may not be found for each item on NRC Form 313. Specific NRC references used in the development of this guidance document are included in Appendix B.

NRC Form 313 does not have sufficient space for applicants to provide full responses to Items 5 through 11; as indicated on the Form, the answers to those items are to be provided on separate sheets of paper and submitted with the completed NRC Form 313. For the convenience of applicants and for streamlined handling of applications for commercial radiopharmacy licenses in the new materials licensing process, use Appendix C to provide supporting information, attach it to NRC Form 313, and submit it to NRC.

Appendix D is a checklist that NRC staff uses to review applications and that applicants may use to check for completeness. Appendix E is a sample commercial radiopharmacy license, containing the conditions most often found on these licenses, although not all licenses will have all conditions. Appendices F through S contain additional information on various radiation safety topics, including model procedures. Appendix T includes a table (Table T.1) of NRC incident notification and reporting requirements applicable to commercial radiopharmacies.

In this document, dose or radiation dose means absorbed dose, dose equivalent, effective dose equivalent (EDE), committed dose equivalent (CDE), committed effective dose equivalent (CEDE), or total effective dose equivalent (TEDE). These terms are defined in 10 CFR Part 20. Rem, and its International System of Units (SI) equivalent, sievert (Sv) (1 rem = 0.01 Sv), are used to describe units of radiation exposure or dose. This is done because 10 CFR Part 20 sets dose limits in terms of rem, not rad or roentgen ®. When the radioactive material emits beta and gamma rays, for practical reasons, we assume that 1 R = 1 rad = 1 rem. For alpha-emitting radioactive material, 1 rad is not equal to 1 rem. Determination of dose equivalent (rem) from absorbed dose (rad) from alpha particles requires the use of an appropriate quality factor (Q) value. These Q values are used to convert absorbed dose (rad) to dose equivalent (rem); Q values for alpha particles are addressed in Tables 1004(b)(1) and (2) in 10 CFR 20.1004.

2   AGREEMENT STATES

Certain states, called Agreement States (see Figure 2.1), have entered into agreements with NRC that give them the authority to license and inspect byproduct, source, or special nuclear materials used or possessed within their borders. Any applicant, other than a Federal agency or Federally recognized Indian tribe, who wishes to possess or use licensed material in one of these Agreement States should contact the responsible officials in that State for guidance on preparing an application; file these applications with State officials, not with NRC.

Locations of NRC Offices and Agreement States

Region IV

Region III

Region I

Region II**

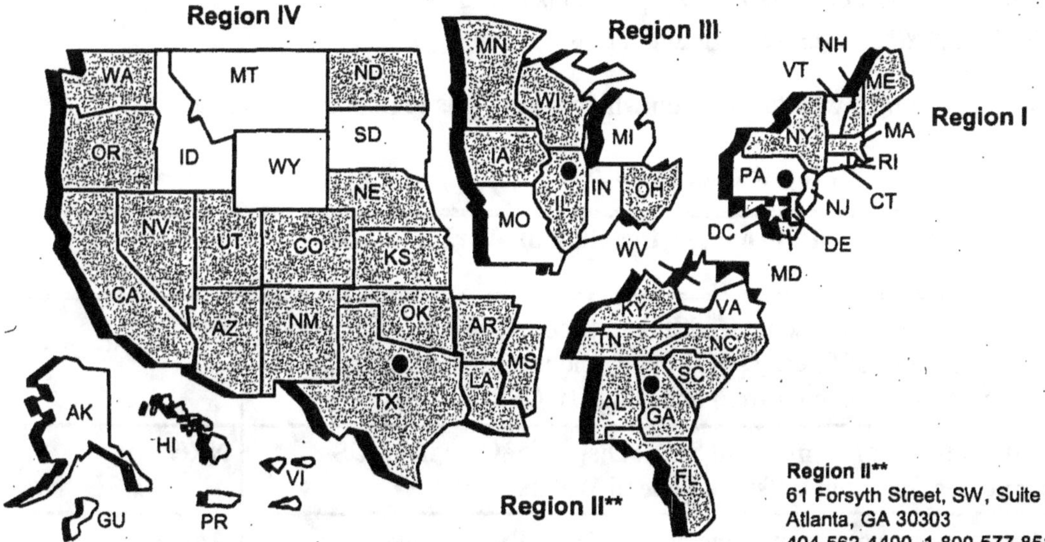

- ● Regional Office   ★ Headquarters
- ▦ 34 Agreement States
- ☐ 16 Non-Agreement States*

Note: Alaska, Hawaii, and Guam are included in Region IV; Puerto Rico and Virgin Islands in Region I

Headquarters
Washington, DC 20555-0001
301-415-7000, 1-800-368-5642

Region I
475 Allendale Road
King of Prussia, PA 19406-1415
610-337-5000, 1-800-432-1156

Region II**
61 Forsyth Street, SW, Suite 23 T85
Atlanta, GA 30303
404-562-4400, 1-800-577-8510

Region III
2443 Warrenville Road, Suite 210
Lisle, IL 60532-4352
630-829-9500, 1-800-522-3025

Region IV
611 Ryan Plaza Drive, Suite 400
Arlington, TX 76011-4005
817-860-8100, 1-800-952-9677

* The 16 Non-Agreement States include three States that have filed letters of intent: Pennsylvania, New Jersey, and Virginia.
** All applicants for materials licenses located in Region II's geographical area must send their applications to Region I.

1556-001q.ppt
053107

Figure 2.1   U.S. Map.   *Location of NRC Offices and Agreement States.*

In the special situation of work at Federally controlled sites in Agreement States, it is necessary to know the jurisdictional status of the land in order to determine whether NRC or the Agreement State has regulatory authority. The NRC has regulatory authority over land determined to be "exclusive Federal jurisdiction," while the Agreement State has jurisdiction over nonexclusive Federal jurisdiction land. Applicants are responsible for finding out, in advance, the jurisdictional status of the specific areas where they plan to conduct licensed operations. The NRC recommends that applicants ask their local contact for the Federal agency controlling the site (e.g., contract officer, base environmental health officer, district office staff) to help determine the jurisdictional status of the land and to provide the information in writing, in order to comply with NRC or Agreement State regulatory requirements, as appropriate. Additional guidance on determining jurisdictional status is found in All Agreement States Letter, SP-96-022, dated February 16, 1996, which is available at http://nrc-stp.ornl.gov/asletters/other/sp96022.pdf.

Table 2.1 provides a quick way to check on which agency has regulatory authority.

Table 2.1   Who Regulates the Activity?

Applicant and Proposed Location of Work	Regulatory Agency
Federal agency or Federally recognized Indian tribe[1] regardless of location (except the Department of Energy [DOE] and, under most circumstances, its prime contractors are exempt from licensing [10 CFR 30.12])	NRC
Non-Federal entity in non-Agreement State, District of Columbia, US territory, or possession, or in Offshore Federal Waters	NRC
Non-Federal entity in Agreement State at non-Federally controlled site	Agreement State
Non-Federal entity in Agreement State at Federally controlled site *not* subject to exclusive Federal jurisdiction	Agreement State
Non-Federal entity in Agreement State at Federally controlled site subject to exclusive Federal jurisdiction	NRC

[1] NRC exercises jurisdiction as the regulatory authority on land where a Federally recognized Indian tribe has tribal jurisdiction. Section 274b Agreements do not give States authority to regulate nuclear material in these areas. Companies owned or operated by native American Indians or non-Indians, wishing to possess or use licensed material in these areas, should contact the appropriate NRC Regional Office to request a license application.

Reference: A current list of Agreement States (including names, addresses, and telephone numbers of responsible officials) is available at the Office of Federal and State Materials and Environmental Management Programs' (FSME) public website, which is located at http://nrc-stp.ornl.gov. As an alternative, request the list from an NRC Regional Office.

3   MANAGEMENT RESPONSIBILITY

The NRC recognizes that effective Radiation Safety Program management is vital to achieving safe and compliant operations. The NRC also believes that consistent compliance with its regulations provides reasonable assurance that licensed activities will be conducted safely and that effective management will result in increased safety and compliance.

> "Management" refers to the processes for conduct and control of a Radiation Safety Program and to the individuals who are responsible for those processes and who have *authority to provide necessary resources* to achieve regulatory compliance.

To ensure adequate management involvement, a duly authorized management representative *must* sign the submitted application acknowledging management's commitments and responsibility for the following:

- Radiation safety, security and control of radioactive materials, and compliance with regulations;

- Completeness and accuracy of the radiation safety records and all information provided to NRC (10 CFR 30.9);

- Knowledge about the contents of the license and application;

- Compliance with current NRC and Department of Transportation (DOT) regulations and the licensee's operating and emergency procedures;

- Commitment to provide adequate resources (including space, equipment, personnel, time, and, if needed, contractors) to the Radiation Protection Program to ensure that the public and workers are protected from radiation hazards and compliance with regulations is maintained;

- Selection and assignment of a qualified individual to serve as the Radiation Safety Officer (RSO) for their licensed activities;

- Prohibition against discrimination of employees engaged in protected activities (10 CFR 30.7);

- Commitment to provide information to employees regarding the employee protection and deliberate misconduct provisions in 10 CFR 30.7 and 10 CFR 30.10, respectively;

- Commitment to obtaining NRC's prior written consent before transferring control of the license; and

- Notification of the appropriate NRC regional administrator in writing, immediately following filing of petition for voluntary or involuntary bankruptcy (10 CFR 30.34(h)).

For information on NRC inspection, investigation, enforcement, and other compliance programs, see the current version of NRC's Enforcement Policy, which is included on NRC's Web site at http://www.nrc.gov/what-we-do/regulatory/enforcement/enforce-pol.html and Inspection Procedure (IP) 87127, "Radiopharmacy Programs," which may be found at http://www.nrc.gov/reading-rm/doc-collections/insp-manual/inspection-procedure/ip87127.pdf.

4 APPLICABLE REGULATIONS

It is the applicant's or licensee's responsibility to obtain up-to-date copies of applicable regulations, read and understand the requirements of each of these regulations, and comply with each applicable regulation. The following Parts of the Code of Federal Regulations (CFR) contain regulations applicable to commercial radiopharmacies:

- 10 CFR Part 2, "Rules of Practice for Domestic Licensing Proceedings and Issuance of Orders";

- 10 CFR Part 19, "Notices, Instructions and Reports to Workers: Inspection and Investigations";

- 10 CFR Part 20, "Standards for Protection Against Radiation";

- 10 CFR Part 21, "Reporting of Defects and Noncompliance";

- 10 CFR Part 30, "Rules of General Applicability to Domestic Licensing of Byproduct Material";

- 10 CFR Part 32, "Specific Domestic Licenses to Manufacture or Transfer Certain Items Containing Byproduct Material";

10 CFR Part 32 allows licensees to prepare radioactive drugs for medical use, provided that the radioactive drug is prepared by either an authorized nuclear pharmacist (ANP) or an individual under the supervision of an ANP as specified in 10 CFR 35.27. In addition, 10 CFR Part 35 specifies the definition of an ANP and medical use in 10 CFR 35.2, and the qualifications of an ANP in 10 CFR 35.55; however, the remaining sections of 10 CFR Part 35 do not apply to commercial radiopharmacy licensees.

- 10 CFR Part 71, "Packaging and Transportation of Radioactive Material";

10 CFR Part 71 requires that licensees or applicants who transport licensed material or who may offer such material to a carrier for transport comply with the applicable requirements of DOT that are found in 49 CFR Parts 107, 171 through 180, and 390 through 397. Copies of DOT regulations can be found at http://hazmat.dot.gov.

- 10 CFR Part 170, "Fees for Facilities, Materials, Import and Export Licenses and Other Regulatory Services Under the Atomic Energy Act of 1954, as Amended"; and

- 10 CFR Part 171, "Annual Fees for Reactor Operating Licenses, and Fuel Cycle Licenses and Materials Licenses, Including Holders of Certificates of Compliance, Registrations, and Quality Assurance Program Approvals and Government Agencies Licensed by the NRC".

Copies of the above documents may be obtained by calling the Government Printing Office (GPO) order desk in Washington, DC at (202) 512-1800, or online at http://www.bookstore.gpo.gov.

A single copy of the above document may be requested from NRC's Regional Offices (see Figure 2.1 for addresses and telephone numbers). In addition, 10 CFR Parts 0-199 can be found on NRC's website at http://www.nrc.gov/reading-rm/doc-collections/cfr. Note that NRC and all other Federal agencies publish amendments to their regulations in the *Federal Register*.

5 HOW TO FILE

5.1 PAPER APPLICATION

Applicants for a materials license should do the following:

- Use the most recent guidance in preparing an application;

- Complete NRC Form 313 (Appendix A) Items 1 through 4, 12, and 13 on the form itself;

- Complete NRC Form 313 Items 5 through 11 on supplementary pages, or use Appendix C;

- Complete NRC Form 313A (ANP) (Appendix G) to document Authorized Nuclear Pharmacist training and experience, if electing to complete this supplemental form;

- Provide sufficient detail for NRC to determine that equipment, facilities, training, experience, and the Radiation Safety Program are adequate to protect health and safety and minimize danger to life and property;

- For each separate sheet, other than NRC Form 313A and Appendix C, that is submitted with the application, identify and cross-reference it to the item number on the application or the topic to which it refers;

- Submit all documents, typed, on 8-1/2 x 11-inch paper;

- Avoid submitting proprietary information unless it is absolutely necessary;

- If submitted, proprietary information and other sensitive information must be clearly identified (see Section 5.2 below);

- Submit an original, signed application and one copy; and

- Retain one copy of the license application for future reference.

Applications must be signed by the applicant, licensee, or a person duly authorized as required by 10 CFR 30.32(c); see Section 8.13, "Certification."

5.2 IDENTIFYING AND PROTECTING SENSITIVE INFORMATION

All licensing applications, except for portions containing sensitive information, will be made available for review in NRC's Public Document Room and electronically at the Public Electronic Reading Room. For more information on the Public Electronic Reading Room, visit www.nrc.gov.

Several types of sensitive information must be identified, marked, and protected against unauthorized disclosure to the public. Key examples are as follows:

- Proprietary Information/Trade Secrets: If it is necessary to submit proprietary information or trade secrets, follow the procedure in 10 CFR 2.390(b). Failure to follow this procedure could result in disclosure of the proprietary information to the public or substantial delays in processing the application.

- Private information: Personal information about employees or other individuals should not be submitted unless specifically requested by NRC. Examples of private information are: social security number, home address, home telephone number, date of birth, and radiation dose information. If private information is submitted, it should be separated from the public portion of the application and clearly marked: "Privacy Act Information - Withhold Under 10 CFR 2.390."

- Security-Related Information: Following the events of September 11, 2001, NRC changed its procedures to avoid release of information that terrorists could use to plan or execute an attack against facilities or citizens in the United States. As a result, certain types of information are no longer routinely released and are treated as sensitive unclassified information. For example, certain information about the quantities and locations of radioactive material at licensed facilities, and associated security measures, are no longer released to the public. Therefore, sensitive security-related information in an application should be marked as specified in Regulatory Issue Summary 2005-31, available at http://www.nrc.gov/reading-rm/doc-collections/gen-comm/reg-issues/2005/ri200531.pdf. Additional information on procedures and any updates are available at http://www.nrc.gov/reading-rm/sensitive-info.html.

5.3 PAPER FORMAT AND ELECTRONIC FORMAT

The NRC's new licensing process will be faster and more efficient, in part, through acceptance and processing of electronic applications at some future date. The NRC will continue to accept paper applications. However, these will be scanned through an optical character reader to convert them to electronic format. To ensure a smooth transition to electronic applications, applicants should:

- Submit printed or typewritten – not handwritten – text on smooth, crisp paper that will feed easily into the scanner;

- Choose typeface designs that are sans serif, such as Arial, Helvetica, Futura, Univers; the text of this document is in a serif font called Times New Roman;

- Use 12-point or larger font;

- Avoid stylized characters such as script, italic, etc.;

- Ensure that the print is clear and sharp;

- Ensure that there is high contrast between the ink and paper (black ink on white paper is best).

As the electronic licensing process develops, it is anticipated that NRC may provide mechanisms for filing applications via CD-ROM and through the Internet. Additional filing instructions will be provided as NRC implements these new mechanisms. When the electronic process becomes available, applicants may file electronically instead of on paper.

6 WHERE TO FILE

Applicants wishing to possess or use licensed material in any State or U.S. territory or possession subject to NRC jurisdiction must file an application with the NRC Regional Office for the locale in which the material will be possessed and/or used. Figure 2.1 shows NRC's four Regional Offices and their respective areas for licensing purposes and identifies Agreement States. Note that all materials applications are submitted to Regions I, III, or IV. All applicants for materials licenses located in Region II's geographical area should send their applications to Region I.

In general, applicants wishing to possess or use licensed material in Agreement States must file an application with the Agreement State, not NRC. However, if work will be conducted at Federally controlled sites in Agreement States, applicants must first determine the jurisdictional status of the land in order to determine whether NRC or the Agreement State has regulatory authority. See the section on "Agreement States" for additional information.

7 LICENSE FEES

Each application for which a fee is specified must be accompanied by the appropriate fee. Refer to 10 CFR 170.31 to determine the amount of the fee. The NRC will not issue the licensing action prior to fee receipt. Consult 10 CFR 170.11 for information on exemptions from these fees. Once technical review has begun, no fees will be refunded; application fees will be charged regardless of NRC's disposition of an application or the withdrawal of an application.

Most NRC licensees are also subject to annual fees; refer to 10 CFR 171.16. Consult 10 CFR 171.11 for information on exemptions from annual fees and 10 CFR 171.16(c) on reduced annual fees for licensees that qualify as "small entities."

Direct all questions about NRC's fees or completion of Item 12 of NRC Form 313 to the Office of the Chief Financial Officer at NRC Headquarters in Rockville, Maryland, (301) 415-7554. Information about fees may also be obtained by calling NRC's toll free number (800) 368-5642, extension 415-7554. The e-mail address is fees@nrc.gov.

8 CONTENTS OF AN APPLICATION

The following comments apply to the indicated items on NRC Form 313 (Appendix A).

All items in the application should be completed in enough detail for NRC to determine that the proposed equipment, facilities, training and experience, and the Radiation Safety Program satisfy regulatory requirements and are adequate to protect health and minimize danger to life and property. Consideration shall be given, when developing the application, to the concepts of keeping exposure as low as reasonably achievable (ALARA) and minimizing contamination.

Regarding ALARA, 10 CFR 20.1101(b) states "The licensee *shall* use, to the extent practicable, procedures and engineering controls based upon sound radiation protection principles to achieve occupational doses and doses to members of the public that are as low as is reasonably achievable (ALARA)." ALARA concepts and philosophy are discussed in Regulatory Guide 8.10, "Operating Philosophy for Maintaining Occupational Radiation Exposures As Low As Is Reasonably Achievable." Applicants for commercial radiopharmacy licenses must address ALARA considerations in all aspects of their programs (e.g., monitoring and controlling external and internal personnel exposure, monitoring and controlling air and liquid effluents). ALARA considerations, including establishing administrative action levels and monitoring programs, should be documented in the application.

Under 10 CFR 20.1406, license applicants are required to describe how facility design and procedures for operation will minimize, to the extent practicable, contamination of the facility and the environment, facilitate eventual decommissioning, and minimize, to the extent practicable, the generation of radioactive waste. Like ALARA, applicants must address these concerns in all aspects of their programs.

All information submitted to NRC during the licensing process will be incorporated as part of the license and will be subject to review during inspection.

8.1 ITEM 1: LICENSE ACTION TYPE

THIS IS AN APPLICATION FOR (Check appropriate item)

Type of Action	License No.
[ ] A. New License	Not Applicable
[ ] B. Amendment	XX-XXXXX-XX
[ ] C. Renewal	XX-XXXXX-XX

Check box A for a new license request.

Check box B for an amendment[1] to an existing license; provide license number.

Check box C for a renewal[1] of an existing license; provide license number.

8.2 ITEM 2: APPLICANT'S NAME AND MAILING ADDRESS

List the legal name of the applicant's corporation or other legal entity with direct control over use of the radioactive material; a division or department within a legal entity may not be a licensee. An individual may be designated as the applicant only if the individual is acting in a private capacity and the use of the radioactive material is not connected with employment in a corporation or other legal entity. Provide the mailing address where correspondence should be sent. A Post Office box number is an acceptable mailing address.

Notify NRC of changes in the mailing address; these changes do not require a fee.

Note: NRC must be notified before control of the license is transferred or when bankruptcy proceedings have been initiated. See below for more details. NRC Information Notice (IN) 97-30, "Control of Licensed Material during Reorganizations, Employee-Management Disagreements, and Financial Crises," dated June 3, 1997, discusses the potential for the security and control of licensed material to be compromised during periods of organizational instability.

Timely Notification of Transfer of Control

Regulation: 10 CFR 30.34(b).

Criteria: Licensees must provide full information and obtain NRC's *prior written consent* before transferring control of the license, or, as some licensees call it, "transferring the license."

---

[1] See "Amendments and Renewals to a License" later in this document.

Discussion: Transferring control may be the result of mergers, buyouts, or majority stock transfers. Although it is not NRC's intent to interfere with the business decisions of licensees, it is necessary for licensees to obtain prior NRC written consent. This is to ensure the following:

- Radioactive materials are possessed, used, or controlled only by persons who have valid NRC licenses;

- Materials are properly handled and secured;

- Persons using these materials are competent and committed to implementing appropriate radiological controls;

- A clear chain of custody is established to identify who is responsible for disposition of records and licensed material; and

- Public health and safety are not compromised by the use of such materials.

Response from Applicant: No response is required from an applicant for a new license. However, current licensees should refer to NUREG-1556, Vol. 15, "Guidance About Changes of Control and About Bankruptcy Involving Byproduct, Source, or Special Nuclear Materials Licenses," dated November 2000, for more information on transfer of ownership.

Notification of Bankruptcy Proceedings

Regulation: 10 CFR 30.34(h).

Criteria: Immediately following the filing of a voluntary or involuntary petition for bankruptcy for or against a licensee, the licensee must notify the appropriate NRC Regional Administrator, in writing, identifying the bankruptcy court in which the petition was filed and the date of filing.

Discussion: Even though a licensee may have filed for bankruptcy, the licensee remains responsible for all regulatory requirements. NRC must know when licensees are in bankruptcy proceedings in order to determine whether all licensed material is accounted for and adequately controlled and whether there are any public health and safety concerns (e.g., contaminated facility). NRC shares the results of its determinations with other involved entities (e.g., trustee), so that health and safety issues can be resolved before bankruptcy actions are completed.

Response from Applicant: None required at the time of application for a new license. Licensees must immediately (within 24 hours) notify NRC following the filing of a voluntary or involuntary petition for bankruptcy for or against the licensee.

Reference: See NUREG-1556, Vol. 15, "Guidance About Changes of Control and About Bankruptcy Involving Byproduct, Source, or Special Nuclear Materials Licenses," dated November 2000.

8.3   ITEM 3: ADDRESS(ES) WHERE LICENSED MATERIAL WILL BE USED OR POSSESSED

Specify the street address, city, and State or other descriptive address (e.g., on Highway 10, 5 miles east of the intersection of Highway 10 and State Route 234, Anytown, State) for each facility. The descriptive address should be sufficient to allow an NRC inspector to find the facility location. Sketches or street maps indicating the nearest intersection and the location of the proposed facility would be helpful but are not required. A Post Office Box address is not acceptable (See Figure 8.1). Documents that give exact location of use should be marked "Security-Related Information – Withhold Under 10 CFR 2.390."

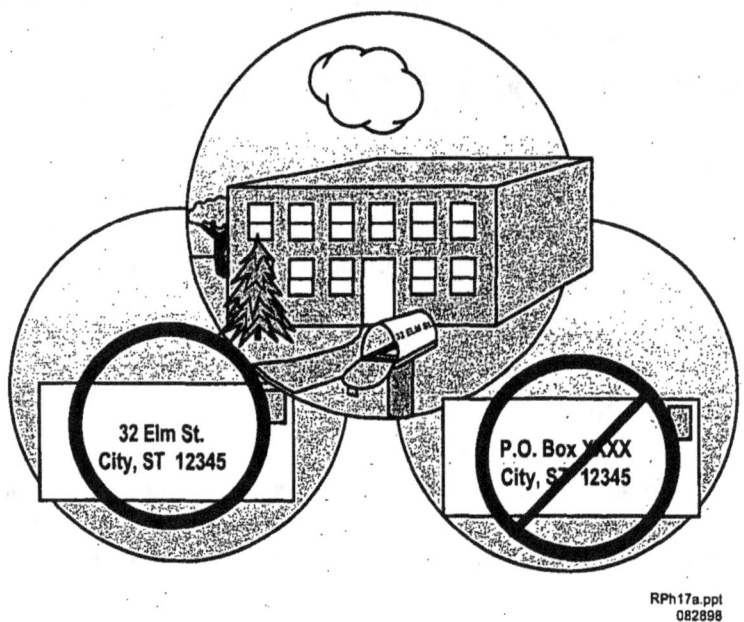

RPh17a.ppt
082898

Figure 8.1 Location of Use. *An acceptable location of use specifies street address, city, State, and zip code and does **not** include a post office box number.*

An NRC-approved license amendment is required before receiving, using, and storing licensed material at an address or location not listed on the license.

Being granted an NRC license does not relieve a licensee from complying with other applicable Federal, State, or local regulations (e.g., local zoning requirements).

Note: As discussed later under "Financial Assurance and Recordkeeping for Decommissioning," licensees must maintain permanent records describing where licensed material was used or stored while the license was in force. This is important for making future determinations about the release of these locations for unrestricted use (e.g., before the license is terminated). Acceptable records are sketches, written descriptions of the specific locations or room numbers where licensed material is used or stored, and any records of spills or other unusual occurrences involving the spread of contamination in or around the licensee's facilities.

8.4 ITEM 4: PERSON TO BE CONTACTED ABOUT THIS APPLICATION

Identify the individual who can answer questions about the application, and include his or her telephone number. This individual, usually the RSO, will serve as the point of contact during the review of the application and during the period of the license. If this individual is not a full-time employee of the licensed entity, his or her position and relationship should be specified. No individual other than the duly authorized applicant may, for any licensing matter, act on behalf of the applicant or provide information without the applicant's written authorization. NRC should be notified if the person assigned to this function changes or if his/her telephone number changes. Notification of a contact change is for information only and would not be considered an application for license amendment, unless the notification involves a change in the contact person who is also the RSO.

As indicated on NRC Form 313 (Appendix A), Items 5 through 11 should be submitted on separate sheets of paper. Applicants may use Appendix C for this purpose and should note that using the suggested wording of responses and committing to using the model procedures in this report will expedite NRC's review.

8.5 ITEM 5: RADIOACTIVE MATERIAL

8.5.1 UNSEALED AND/OR SEALED BYPRODUCT MATERIAL

Regulations: 10 CFR 30.4, 10 CFR 30.33, 10 CFR 32.72(a)(3), 10 CFR 30.32(g), 10 CFR 32.210, 10 CFR 30.32(i).

Criteria: Applicants must submit information specifying each radionuclide requested, the form, and the maximum activity to be possessed at any one time. For sealed sources, the applicant must also submit the manufacturer and model number of each requested sealed source.

On August 8, 2005, the EPAct gave NRC regulatory authority over new byproduct material. This new byproduct material includes accelerator-produced radionuclides and naturally occurring materials such as discrete sources of radium-226. See 10 CFR 30.4 for a complete definition of the term "byproduct material."

Applicants and licensees should determine whether they possess or will possess sealed sources or devices containing this new byproduct material, which would include check, calibration, and reference sources, that are not generally licensed or exempt from licensing Applicants will have to request authorization for possession of these sealed source(s) or device(s).

Note that the naturally occurring and accelerator-produced sealed sources and devices that were manufactured prior to November 30, 2007, may not have been registered by the NRC in accordance with 10 CFR 32.210(c) or an Agreement State. If the applicant possesses unregistered sources and/or devices and is unable to provide all categories of information specified in 10 CFR 32.210(c), the applicant must submit the information required by 10 CFR 30.32(g)(3).

Discussion: Each authorized radionuclide is listed on an NRC license by its element name, form, and the maximum amount the licensee may possess at any one time (maximum possession limit), as shown in items 6, 7, and 8 of the sample licenses in Appendix E.

The applicant should list each requested radionuclide by its element name and its mass number (e.g., technetium-99m, indium-111, and fluorine-18) in Item 5. Note in the sample license in Appendix E that NRC provides broad authorization to permit radiopharmacy licensees flexibility to prepare and distribute a range of radionuclides as new radioactive drugs are developed. It is necessary to specify whether the material will be acquired and used in unsealed or sealed form or in the case of radium-226, in the form of a discrete source. The name of the specific chemical compound that contains the radionuclide is not generally required.

For unsealed radioactive material, it is also necessary to specify whether requested radionuclides will be handled in volatile or nonvolatile form, since additional safety precautions are required when handling and using material in a volatile form. For example, when requesting authorization to possess and distribute iodine-131, the applicant must specify whether the material will be manipulated at the radiopharmacy in a volatile form (e.g., compounding of iodine-131 capsules) or received in the form in which it will be distributed (e.g., redistribution of sealed, unopened vials of iodine-123). Also, if the pharmacy possesses discrete sources of radium-226, the discrete source should be described, since additional precautions may need to be taken if the source is compromised. Applicants requesting discrete sources of radium-226 and authorization to manipulate volatile radioactive material must describe appropriate facilities and engineering controls in response to Section 8.9, "Facilities and Equipment," and radiation safety procedures for handling of such material in specific responses to Section 8.10.4, "Occupational Dose"; Section 8.10.5, "Public Dose"; Section 8.10.6, "Safe Use of Radionuclides and Emergency Procedures"; and Section 8.10.7, "Surveys."

The anticipated possession limit in becquerels (Bq) or curies (Ci) for each radionuclide should also be specified. Possession limits must include the total anticipated inventory, including licensed material in storage and waste, and should be commensurate with the applicant's needs and facilities for safe handling. Applicants should review the requirements for submitting a certification for financial assurance for decommissioning before specifying possession limits of any radionuclide with a half-life greater than 120 days. These requirements are discussed in the Section on Financial Assurance and Decommissioning.

Applicants that produce radionuclides using an accelerator (e.g., PET cyclotron) would list only those radionuclides produced for use in the pharmacy (e.g., fluorine-18). All other radionuclides associated with PET radionuclide production (e.g., activation products) should be provided with the application submitted in accordance with NUREG-1556, Vol. 21, "Consolidated Guidance About Materials Licenses: Program-Specific Guidance About Possession License for Production of Radioactive Materials Using an Accelerator."

A safety evaluation of sealed sources and devices is performed by NRC or an Agreement State before authorizing a manufacturer (or distributor) to distribute them to specific licensees. The safety evaluation is documented in a Sealed Source and Device Registry (SSDR) certificate. Applicants must provide the manufacturer's name and model number for each requested sealed source and device, so that NRC can verify that they have been evaluated in an SSDR certificate or specifically approved on a license.

A pharmacy possessing a sealed source containing the new byproduct material as defined by the EPAct that does not have an SSDR certificate must provide the information required under 10 CFR 30.32(g). A pharmacy that intends to manufacture, distribute, or redistribute such a source will need to request a safety evaluation by NRC or an Agreement State.

Consult with the proposed supplier, manufacturer, or distributor to ensure that requested sources and devices are compatible with and conform to the sealed source and device designations registered with NRC or an Agreement State. Licensees may not make any changes to the sealed source, device, or source/device combination that would alter the description or specifications from those indicated in the respective SSDR certificates, without obtaining NRC's prior permission in a license amendment. To ensure that applicants use sources and devices according to the certificates, they may want to get a copy of the certificate and review it or discuss it with the manufacturer.

To obtain copies of the SSDR certificate, applicants should contact the manufacturer or distributor of the device or contact NRC's Office of Federal and State Materials and Environmental Management Programs. For additional guidance relating to sealed sources and devices, see also NUREG-1556, Vol. 3., Rev. 1, "Applications for Sealed Source and Device Evaluation and Registration."

The applicant must also request authorization to possess depleted uranium if it will be used as shielding for molybdenum-99/technetium-99m generators. Depleted uranium is frequently used as shielding for generators when the molybdenum-99 activity is greater than 148 gigabecquerels (4 curies). In 10 CFR 40.13(c)(6), depleted uranium is exempt from the requirements for a license to the extent that the material is used as a shipping container, such as when molybdenum-99/ technetium-99m generators are in transit from their manufacturer to the pharmacy; however, a specific license or authorization from NRC is needed to possess and use the depleted uranium as a shield during the time that the pharmacy uses or stores the generator at its facility. The applicant must specify the total amount of depleted uranium, in kilograms, that will be needed.

If an applicant requests quantities of licensed material in excess of those specified in 10 CFR 30.72, "Schedule C - Quantities of Radioactive Materials Requiring Consideration of the Need for an Emergency Plan for Responding to a Release," the applicant must either submit an emergency plan for responding to a release of radioactive materials or perform an evaluation showing that the maximum dose to a person offsite due to a release of radioactive materials would not exceed 10 millisieverts (mSv) (1 rem) effective dose equivalent or 50 mSv (5 rems) to the thyroid. For radiopharmacies, iodine-131 is the radionuclide most likely to trigger the need for a emergency plan due to its Schedule C quantity of 10 curies.

Licensees must submit a license amendment and receive NRC authorization before they may make changes in the types, forms, and quantities of materials possessed.

Response from Applicant:

- For unsealed materials:

  — Identify each radionuclide (element name and mass number) that will be used, the form, and the maximum requested possession limit.

AND

- For potentially volatile materials (e.g., iodine-123, iodine-131):

  — Specify whether open containers of the materials will be manipulated at the radiopharmacy.

- For sealed sources and discrete sources of radium-226:

  — Identify each radionuclide (element name and mass number) that will be used in each source;

  — Provide the manufacturer's (distributor's) name and model number for each sealed source and device and discrete source of radium-226 requested;

  — Confirm that each sealed source, device, source/device combination, and discrete source of radium-226 is registered as an approved sealed source, device, or discrete source by the NRC or an Agreement State;

  — Confirm that the activity per source and maximum activity in each device will not exceed the maximum activity listed on the approved certificate of registration issued by NRC or by an Agreement State; and

  — If the above information cannot be provided for the discrete source of radium-226, describe the discrete source.

- For depleted uranium, specify the total amount (in kilograms).

8.5.2   FINANCIAL ASSURANCE AND RECORDKEEPING FOR DECOMMISSIONING

Regulations: 10 CFR 30.35, 10 CFR 30.34(b).

Criteria: A licensee authorized to possess radioactive material in excess of the limits specified in 10 CFR 30.35 must submit a decommissioning funding plan (DFP) or provide a certification of financial assurance (FA) for decommissioning. Even if a DFP or FA is not required, licensees are required to maintain, in an identified location, decommissioning records related to structures and equipment where radioactive materials are used or stored and related to leaking sources. Pursuant to 10 CFR 30.35(g), licensees must transfer records important to decommissioning to either of the following:

- The new licensee before licensed activities are transferred or assigned according to 10 CFR 30.34(b); or

- The appropriate NRC Regional Office before the license is terminated.

Discussion: The requirements for financial assurance are specific to the types and quantities of byproduct material authorized on a license. Most commercial radiopharmacy applicants and licensees do not need to take any action to comply with the financial assurance requirements, because the vast majority of radioactive materials they possess and redistribute do not have half-lives greater than 120 days and the total inventory of licensed materials with half-lives greater than 120 days does not exceed the thresholds in 10 CFR 30.35(b) and (d).

Applicants requesting more than one radionuclide may determine whether financial assurance for decommissioning is required by calculating, for each radionuclide with a half-life greater than 120 days possessed, the ratio between the activity possessed, in curies, and the radionuclide's threshold activity requiring financial assurance, in curies. If the sum of such ratios for all of the radionuclides possessed exceeds "1" (i.e., "unity"), applicants must submit evidence of financial assurance for decommissioning.

The same regulation also requires that licensees maintain records important to decommissioning in an identified location. All commercial nuclear pharmacy licensees need to maintain records of structures and equipment where radioactive material was used or stored. As-built drawings with modifications of structures and equipment shown as appropriate fulfill this requirement. If drawings are not available, licensees shall substitute appropriate records (e.g., a sketch of the room or building or a narrative description of the area) concerning the specific areas and locations. If no records exist regarding structures and equipment where radioactive materials were used or stored, licensees shall make all reasonable efforts to create such records based upon historical information (e.g., employee recollections). In addition, if radiopharmacy licensees have experienced unusual occurrences (e.g., incidents that involve spread of contamination, leaking sources), they should also maintain records about contamination that remains after cleanup or that may have spread to inaccessible areas.

For radiopharmacy licensees whose contamination incidents did not involve radioactive materials with half-lives exceeding 120 days and whose sealed sources have never leaked, acceptable records important to decommissioning are sketches or written descriptions of the specific locations where radioactive material was used or stored.

Response from Applicant: No response is needed from most applicants. If financial assurance is required, submit the documentation required under 10 CFR 30.35. NUREG-1757, Vol. 3, "Consolidated NMSS Decommissioning Guidance: Financial Assurance, Recordkeeping, and Timeliness," dated September 2003, contains approved wording for each of the mechanisms authorized by the regulation to guarantee or secure funds.

Licensees must transfer records important to decommissioning either to the new licensee before licensed activities are transferred or assigned in accordance with 10 CFR 30.34(b) or to the appropriate NRC Regional Office before the license is terminated.

References: See NUREG-1757, Vol. 3, "Consolidated NMSS Decommissioning Guidance: Financial Assurance, Recordkeeping, and Timeliness," dated September 2003.

8.6   ITEM 6: PURPOSE(S) FOR WHICH LICENSED MATERIAL WILL BE USED

The distribution of radioactive materials by commercial radiopharmacies is authorized by several distinct regulations. The appropriate regulation to refer to depends on the nature of the material, the purpose(s) for which it will be used, and to whom it is sent. See Figure 8.2 and narrative description below.

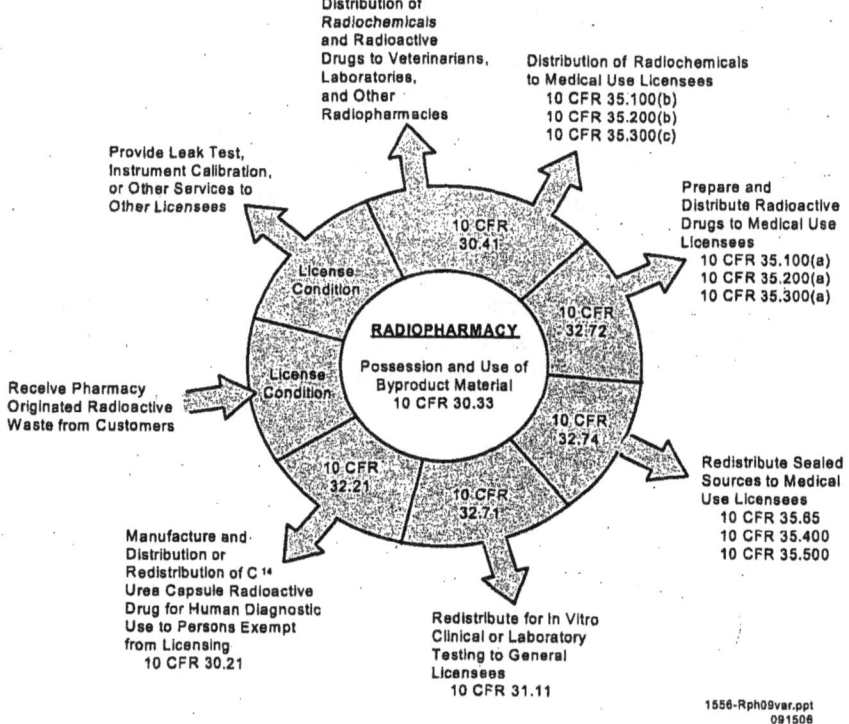

Figure 8.2   Purpose Wheel.

Figure 8.2 Description	
Activities	**Authorized By**
Provide Leak Test, Instrument Calibration, or Other Services to Other Licensees	License Condition
Distribution of Radiochemicals and Radioactive Drugs to Veterinarians, Laboratories, and Other Radiopharmacies	10 CFR 30.41
Distribution of Radiochemicals to Medical Use Licensees: 10 CFR 35.100(b), 10 CFR 35.200(b), 10 CFR 35.300(c)	
Prepare and Distribute Radioactive Drugs to Medical Use Licensees: 10 CFR 35.100(a), 10 CFR 35.200(a), 10 CFR 35.300(a)	10 CFR 32.72
Redistribute Sealed Sources to Medical Use Licensees: 10 CFR 35.65, 10 CFR 35.400, 10 CFR 35.500	10 CFR 32.74
Redistribute for *In Vitro* Clinical or Laboratory Testing to General Licensees: 10 CFR 31.11	10 CFR 32.71
Manufacture and Distribution or Redistribution of C^{14} Urea Capsule Radioactive Drug for Human Diagnostic Use to Persons Exempt from Licensing: 10 CFR 30.21	10 CFR 32.21
Receive Pharmacy-Originated Radioactive Waste from Customers	License Condition

8.6.1 DISTRIBUTION AND REDISTRIBUTION OF SEALED AND UNSEALED MATERIALS

Regulations: 10 CFR 30.41, 10 CFR 32.71, 10 CFR 32.72, and 10 CFR 32.74.

Criteria: The applicant must specify the radioactive material it intends to distribute and redistribute.

Discussion: Radiochemicals are those materials that either require further manipulation to be suitable for human use or are not intended for human use. Examples include raw materials received from a non-10 CFR 32.72 supplier (chemical grade materials). Radioactive drugs are those materials suitable for human use and include radiobiologics (e.g., monoclonal antibodies and technetium-99m-tagged red blood cells) and radiopharmaceuticals. However, the terms, "radiopharmaceutical" and "radioactive drug" will be used interchangeably in this guidance document, and reference to one is not meant to exclude the other.

Distribution activities are normally classified as either "distribution" or "redistribution." "Distribution" applies to those radioactive drugs and radiochemicals initially prepared by the pharmacy. "Redistribution" refers to those materials received from another person, authorized pursuant to either 10 CFR 32.71, 10 CFR 32.72, or 10 CFR 32.74, depending on the product distributed (i.e., *in vitro* kits, other radiopharmaceuticals, or sealed sources for medical use, respectively).

The distribution of radioactive materials to other persons requires specific approval from NRC, either by NRC regulation or by a license authorizing the activity. The initial distribution of radioactive drugs for medical use must be prepared by a person licensed pursuant to 10 CFR 32.72. The redistribution of *in vitro* kits and sealed sources containing byproduct material for medical use is authorized pursuant to 10 CFR 32.71 and 10 CFR 32.74, respectively, provided that the materials are not repackaged and the labels are not altered. The *in vitro* kits and sealed sources for medical use intended for redistribution must be initially distributed by a person licensed pursuant to 10 CFR 32.71 or 10 CFR 32.74, respectively. The transfer of radioactive materials for nonmedical use, including radiochemicals, and sealed calibration and reference sources, is authorized pursuant to 10 CFR 30.41.

All radioactive material listed above shall be distributed only to persons authorized by an NRC or Agreement State license to receive such materials, or by a general license (10 CFR 31.11, or equivalent Agreement State regulation) to receive *in vitro* test materials.

Initial distribution of unsealed byproduct material in the form of radiopharmaceuticals intended for human diagnostic and therapeutic use by medical licensees comprises the bulk of virtually all radiopharmacy activities. Prior to the transfer, distribution, or redistribution of any licensed material, the radiopharmacy must verify that the transferee's license authorizes the receipt of the type, form, and quantity of byproduct material to be transferred. The pharmacy should verify that the address to which radioactive materials are delivered is an authorized location of use listed on the customer's license. Five methods that can be used to meet the license verification requirement are listed in 10 CFR 30.41(d). The most common form of verification is for the radiopharmacy to possess a valid copy of the customer's NRC or Agreement State license or other applicable document (e.g., *in vitro* registration certificate/NRC Form 483).

Response From Applicant: Provide the following, as applicable:

For radiopharmaceuticals:

- Confirm that radiopharmaceuticals will be prepared under the supervision of an ANP or will be obtained from a supplier authorized pursuant to 10 CFR 32.72, or under equivalent Agreement State requirements; and

- Describe all licensed material to be distributed or redistributed.

For generators:

- Confirm that the generators will be obtained from a manufacturer licensed pursuant to 10 CFR 32.72, or under equivalent Agreement State requirements; and

- Confirm that unused generators will be redistributed without opening or altering the manufacturer's packaging.

For redistribution of used generators:

- Describe the procedures and instructions for safely repackaging the generators, including the use of the manufacturer's original packaging and minimization of migration of radioactive fluids out of the generator during transport;

- Confirm that the manufacturer's packaging and labeling will not be altered;

- Confirm that the generator will not be distributed beyond the expiration date shown on the generator label;

- Confirm that the redistributed generator will be accompanied by the manufacturer-supplied leaflet or brochure that provides radiation safety instructions for handling and using the generator; and

- Confirm that only generators used in accordance with the manufacturer's instructions will be redistributed.

Note: Although redistribution of used generators may be authorized by NRC, NRC approval does not relieve the licensee from complying with applicable Food and Drug Administration (FDA) or other Federal and State requirements.

For redistribution of sealed sources — for brachytherapy or diagnosis:

- Confirm that the sealed sources for brachytherapy or diagnosis to be redistributed will be obtained from a manufacturer authorized to distribute sealed sources for brachytherapy or diagnosis in accordance with a specific license issued pursuant to 10 CFR 32.74, or under equivalent Agreement State requirements; and

- Confirm that the manufacturer's packaging, labeling, and shielding will not be altered and that redistributed sources will be accompanied by the manufacturer-supplied package insert, leaflet, brochure, or other document that provides radiation safety instructions for handling and storing the sources.

For redistribution of calibration and reference sealed sources:

- Confirm that calibration and reference sealed sources to be redistributed to medical use licensees will be obtained from a person licensed pursuant to 10 CFR 32.74 or under equivalent Agreement State requirements, to initially distribute such sources; and

- Confirm that the manufacturer's labeling and packaging will not be altered and that redistributed sources will be accompanied by the manufacturer-supplied calibration certificate and the leaflet, brochure, or other document that provides radiation safety instructions for handling and storing the sources.

For redistribution of prepackaged units for *in vitro* tests:

- Confirm that the prepackaged units for *in vitro* tests to be redistributed will have been obtained from a manufacturer authorized to distribute the prepackaged units for *in vitro* tests in accordance with a specific license issued pursuant to 10 CFR 32.71, or under an equivalent license of an Agreement State.

For redistribution to general licensees:

- Confirm that the manufacturer's packaging and labeling of the prepackaged units for *in vitro* tests will not be altered in any way; and

- Confirm that each redistributed prepackaged unit for *in vitro* tests will be accompanied by the manufacturer-supplied package insert, leaflet, or brochure that provides radiation safety instructions for general licensees.

For redistribution to specific licensees:

- Confirm that the labels, package insert, leaflet, brochure, or other documents accompanying the redistributed prepackaged units for *in vitro* tests will NOT reference general licenses, exempt quantities, or NRC's regulations that authorize a general license (e.g., 10 CFR 31.11); and

- Confirm that the labeling on redistributed prepackaged units for *in vitro* tests will conform to the requirements of 10 CFR 20.1901 and 20.1904.

For redistribution of discrete sources of radium-226:

- Confirm that the discrete sources of radium-226 will be obtained by a manufacturer authorized to distribute it.

- Confirm that the manufacturer's packaging, labeling, and shielding will not be altered and that redistributed sources will be accompanied by the manufacturer-supplied package insert, leaflet, brochure, or other document that provides radiation safety instructions for handling and storing sources.

- If the above cannot be confirmed, contact the appropriate NRC Regional Office for assistance.

8.6.2 PREPARATION OF RADIOPHARMACEUTICALS

Regulation: 10 CFR 32.72(b).

Criteria: The preparation of radiopharmaceuticals for commercial distribution to medical users requires specific authorization.

Discussion: The bulk of radiopharmacy activities involves the preparation of radiopharmaceuticals for commercial distribution to medical users.

Response From Applicant: The applicant should indicate the types of radiopharmaceutical preparation activities it intends to perform (e.g., compounding of iodine-131 capsules, radioiodination, chemical synthesis of PET radiopharmaceuticals, and technetium-99m kit preparation).

8.6.3 SEALED SOURCES FOR CALIBRATION AND CHECKS AND POSSESSION OF DISCRETE SOURCES OF RADIUM-226 AND DEPLETED URANIUM

Regulation: 10 CFR 30.33, 10 CFR 30.32(g), 10 CFR 32.210.

Criteria: The applicant must specify the uses for discrete sources of radium-226, sealed sources for reference and calibration, and depleted uranium for shielding.

Discussion: The applicant should describe the intended use of discrete sources of radium-226 and sealed sources. This will normally be for calibration and checks performed only on the applicant's instruments and equipment. Any sources intended for use in a specific instrument calibration device should be identified, along with the manufacturer and model number of the device. The use of depleted uranium for shielding, (e.g., incorporated into molybdenum-99/technetium-99m generators) should also be specified, if appropriate.

Response from Applicant: Supply specific information concerning the use of discrete sources of radium-226, sealed sources for reference and calibration, and depleted uranium for shielding.

8.6.4 SERVICE ACTIVITIES

Regulation: 10 CFR 30.33(a)(1).

Criteria: The applicant must specify the radiation protection services it intends to provide to other licensees (e.g., customers), if the service involves the applicant's possession of licensed material (e.g., calibration sources and leak test samples).

Discussion: If the applicant intends to provide radiation protection services to customers, the services must be described. Typically these services include instrument calibration and sealed source leak testing. Specific guidance regarding requests to provide service activities is included in NUREG-1556, Vol. 18, "Program-Specific Guidance About Service Provider Licenses," dated November 2000.

Response from Applicant: Specify the customer radiation protection services involving licensed material that will be provided. The applicant should submit specific procedures for all service activities that it intends to provide.

8.7 ITEM 7: INDIVIDUAL(S) RESPONSIBLE FOR RADIATION SAFETY PROGRAM AND THEIR TRAINING EXPERIENCE

Regulations: 10 CFR 30.33(a)(3).

Criteria: The RSO, Authorized Users (AUs), and Authorized Nuclear Pharmacists (ANPs) must have adequate training and experience.

Discussion: Individuals responsible for the Radiation Protection Program are licensee senior management, the RSO, ANPs, and AUs. NRC requires that an applicant be qualified by training and experience to use licensed materials for the purposes requested in such a manner as to protect health and minimize danger to life or property. Specific criteria are given in 10 CFR 35.55(b) and 10 CFR 32.72(b) for acceptable training and experience for ANPs. The minimum training and experience criteria for RSOs and AUs, although not specifically described in NRC's regulations for radiopharmacy licensees, should include a Bachelor's degree in a physical science, or equivalent, and previous experience handling and supervising similar activities. Applicants should note that a résumé or a curriculum vitae does not usually supply all the information needed to evaluate an individual's training and experience.

NRC holds the licensee responsible for the Radiation Protection Program; therefore, it is essential that strong management controls and oversight exist to ensure that licensed activities are conducted properly. Management responsibility and liability are sometimes underemphasized or not addressed in applications and are often poorly understood by licensee employees and managers. Senior management should delegate to the RSO, in writing, sufficient authority, organizational freedom, and management prerogative to communicate with and direct personnel regarding NRC regulations and license provisions and to terminate unsafe activities involving byproduct material. The licensee maintains the ultimate responsibility, nevertheless, for the conduct of licensed activities.

Response from Applicant: Refer to the subsequent sections specific to the individuals described above. Applicants should submit an organizational chart describing the management structure, reporting paths, and the flow of authority between executive management and the RSO.

8.7.1 RADIATION SAFETY OFFICER (RSO)

Regulations: 10 CFR 30.33(a)(3).

Criteria: Each licensee must appoint a qualified individual to act as the RSO. The RSO must have adequate training and experience.

Discussion: NRC requires the name, training, and experience of the proposed RSO to ensure that the applicant has identified a responsible, qualified person to oversee the Radiation Safety Program. When selecting an RSO, the applicant should keep in mind the duties and responsibilities of the position, and select an individual who is qualified and has the time and resources to fulfill those duties and responsibilities. Typical duties and responsibilities of a radiopharmacy RSO are included in Appendix H.

The RSO needs a level of basic technical knowledge sufficient to understand the work to be performed with byproduct materials at the radiopharmacy and to be qualified by training and experience to perform the duties required for that position. Any individual who has sufficient training and experience to be named as an ANP is also considered qualified to serve as the facility RSO. The same is true for an AU who has had adequate training and experience in the radiation safety aspects associated with the use of similar types of byproduct material.

The training and experience requirements for the RSO may be met by any of the following:

- Qualification as an ANP;

- Identification as an AU on the license and experience in the use of the types and quantities of licensed material for which the individual has RSO responsibilities; and

- Didactic and work experience.

In order to demonstrate adequate training and experience, the RSO should have (1) as a minimum, a college degree at the Bachelor level, or equivalent training and experience in physical, chemical, biological sciences, or engineering; and (2) training and experience commensurate with the scope of proposed activities. Training should include the following subjects:

- Radiation protection principles;

- Characteristics of ionizing radiation;

- Units of radiation dose and quantities;

- Radiation detection and measurement instrumentation;

- Biological hazards of exposure to radiation (appropriate to types and forms of byproduct material to be used);

- NRC regulatory requirements and standards; and

- Hands-on use of radioactive materials commensurate with the uses proposed by the applicant.

The length of training and experience will depend upon the type, form, quantity, and proposed use of the licensed material requested. The proposed RSO's training and experience should be sufficient to identify and control the anticipated radiation hazards. The requisite training may be obtained from formal courses consisting of lectures and laboratories designed for RSOs presented by academic institutions, commercial radiation safety consulting companies, or appropriate professional organizations. Each hour of training may be counted only once and should be allocated to the most representative topic.

On-the-job training may not be counted toward the hours documenting length of training unless it was obtained as part of a formal training course. A "formal" training course is one that incorporates the following elements:

- A detailed description of the content of the course is maintained on file at the sponsoring institution and can be made available to NRC upon request;

- Evidence that the sponsoring institution has examined the student's knowledge of the course content, is maintained on file at the institution, and can be made available to NRC upon request. This evidence of the student's overall competency in the course material should include a final grade or percentile; and

- A permanent record that the student successfully completed the course is kept at the institution.

The qualifications described above only apply to an RSO for a radiopharmacy that prepares radioactive drugs or redistributes other products. NUREG-1556, Vol. 21, "Consolidated Guidance About Materials Licenses: Program-Specific Guidance About Possession License for Production of Radioactive Materials Using an Accelerator," provides training and experience guidance for individuals that will be RSOs at radionuclide production facilities.

Response from Applicant: Provide the following:

- Name of the proposed RSO;

AND

- A copy of the license (NRC or Agreement State) that authorized the uses requested and on which the individual was specifically named as the RSO, ANP, or AU;

OR

- Description of the training and experience demonstrating that the proposed RSO is qualified by training and experience applicable to commercial nuclear pharmacies.

Note: See Tables G.1 and G.2 in Appendix G for convenient formats to use for documenting hours of training in basic radionuclide handling techniques and hours of experience using radionuclides.

8.7.2 AUTHORIZED NUCLEAR PHARMACIST (ANP)

Regulations: 10 CFR 32.72 (b)(2), (4), and (5); 10 CFR 35.2;
10 CFR 35.55(a) and (b); and 10 CFR 35.59.

Criteria: The ANP must be a State-licensed or State-registered pharmacist with adequate training and experience.

Discussion: Each commercial nuclear pharmacy must have an ANP to prepare or supervise the preparation of radioactive drugs for medical use. An individual who is not qualified to be an ANP may work under the supervision of an ANP.

The criteria for a pharmacist to work as an ANP at a commercial radiopharmacy are described in 10 CFR 32.72(b)(2) and (4). This section of the regulation refers to the training for an ANP, which includes the definition of an ANP in 10 CFR 35.2 (which in turn includes the board certification requirements in 10 CFR 35.55(a)); the training and experience criteria for the alternate pathway described in 10 CFR 35.55(b); and the recentness of training criteria in 10 CFR 35.59 that requires the successful completion of training within 7 years preceding the date of the application. Additional training and experience may be necessary if the time interval is greater than 7 years. Applicants may find it convenient to present this documentation using NRC Form 313A (ANP) in Appendix G. Each hour of training may be listed only once, (i.e., under the most applicable category). The recentness of training requirements apply to board certification as well as to other recognized training pathways.

In implementing the EPAct, NRC "grandfathered" nuclear pharmacists by permitting the licensee to designate a pharmacist as an ANP, if the pharmacist used only accelerator-produced radioactive materials, discrete sources of Ra-226, or both, in the practice of nuclear pharmacy for the uses performed before November 30, 2007, or under the NRC waiver of August 31, 2005. These individuals do not have to meet the requirements of 10 CFR 32.72(b)(2)(i) or (ii). However, the applicant must document that the individual meets the criteria in 10 CFR 32.72(b)(4).

On-the-job training may not be counted toward the hours listed above unless it was obtained as part of a formal training course. A "formal" training course is one that incorporates the following elements:

- A detailed description of the content of the course is maintained on file at the sponsoring institution and can be made available to NRC upon request;

- Evidence that the sponsoring institution has examined the student's knowledge of the course content is maintained on file at the institution and can be made available to NRC upon request. This evidence of the student's overall competency in the course material should include a final grade or percentile; and

- A permanent record that the student successfully completed the course is kept at the institution.

Response from Applicant: For each proposed ANP, provide the following:

- Name of the proposed ANP,

AND

- Pharmacist's license number and issuing entity.

AND

For an individual previously identified as an ANP on an NRC or Agreement State license or permit or by a commercial nuclear pharmacy that has been authorized to identify ANPs (10 CFR 32.72(b)(2)(i)):

• Previous license number (if issued by NRC) or a copy of the license (if issued by an Agreement State) or a copy of a permit issued by an NRC master materials licensee, a permit issued by an NRC or Agreement State broad-scope licensee, or a permit issued by an NRC Master Material License broad-scope permittee on which the individual was named an ANP or a copy of an authorization as an ANP from a commercial nuclear pharmacy that has been authorized to identify ANPs,

OR

For an individual qualifying under 10 CFR 32.72(b)(4):

• Documentation that the individual was a nuclear pharmacist preparing only radioactive drugs containing accelerator-produced radioactive material,

AND

• Documentation that the individual practiced at a pharmacy, a Government agency or Federally recognized Indian Tribe before November 30, 2007, or at all other pharmacies before August 8, 2009, or an earlier date as noticed by the NRC,

OR

For an individual qualifying under 10 CFR 35.55(a):

• Copy of the certification(s) of the specialty board whose certification process has been recognized[2] under 10 CFR 35.55(a),

AND

• Written attestation, signed by a preceptor ANP, that training and experience required for certification have been satisfactorily completed and that a level of competency sufficient to function independently as an ANP has been achieved,

AND

• If applicable, description of recent related continuing education and experience as required by 10 CFR 35.59,

OR

For an individual qualifying under 10 CFR 32.72(b)(2)(ii):

• Description of the training and experience specified in 10 CFR 35.55(b) demonstrating that the proposed ANP is qualified by training and experience,

AND

---

[2] The names of board certifications that have been recognized by NRC or an Agreement State are posted on NRC's web page http://www.nrc.gov/materials/miau/med-use-toolkit.html.

- Written attestation, signed by a preceptor ANP, that training and experience required for certification have been satisfactorily completed and that a level of competency sufficient to function independently as an ANP has been achieved,

AND

- If applicable, description of recent related continuing education and experience as required by 10 CFR 35.59.

Notes:

- NRC Form 313A (ANP), "Authorized Nuclear Pharmacist Training and Experience and Preceptor Attestation [10 CFR 35.55]," may be used to document training and experience for those individuals qualifying under 10 CFR 35.55(a) or (b).

- Descriptions of training and experience will be reviewed using the criteria listed above. The NRC will review the documentation to determine if the applicable criteria in 10 CFR 32.72(b)(2) are met. If the training and experience do not appear to meet the criteria in Subpart B, the NRC may request additional information from the applicant or may request the assistance of the Advisory Committee on the Medical Uses of Isotopes (ACMUI) in evaluating such training and experience.

8.7.3   AUTHORIZED USERS (AU)

Regulation: 10 CFR 30.33(a)(3).

Criteria: Authorized users (AUs) must have adequate training and experience with the types and quantities of licensed material that they propose to use.

Discussion: If the applicant intends to perform functions other than the preparation and distribution of radioactive drugs, the applicant may request that an individual other than an ANP perform and/or supervise those functions. This individual, if approved, would be designated on the license as an AU. These other functions may include leak testing of sealed sources or instrument calibration services for the pharmacy and its customers; however, the term Authorized User, as used in this document, should not be confused with the definition of an "Authorized User" contained in 10 CFR 35.2 for medical use.

In order to demonstrate adequate training and experience, the proposed AU should have (1) as a minimum, a college degree at the Bachelor level, or equivalent training and experience in physical, chemical, biological sciences, or engineering; and (2) training and experience commensurate with the scope of proposed activities. Training should include the following subjects:

- Radiation protection principles,

- Characteristics of ionizing radiation,

- Units of radiation dose and quantities,

- Radiation detection and measurement instrumentation,

- Biological hazards of exposure to radiation (appropriate to types and forms of byproduct material to be used),

- NRC regulatory requirements and standards, and

- Hands-on use of radioactive materials commensurate with uses proposed by the applicant.

The length of training and experience listed above will depend upon the type, form, quantity, and proposed use of the licensed material requested. The proposed AU's training and experience should be sufficient to identify and control the anticipated radiation hazards. The above training may be obtained from formal radiation safety courses consisting of lectures and laboratories presented by academic institutions, commercial radiation safety consulting companies, or appropriate professional organizations. Each hour of training may be counted only once and should be allocated to the most representative topic.

On-the-job training may not count toward the hours listed above unless it was obtained as part of a formal training course. A "formal" training course is one that incorporates the following elements:

- A detailed description of the content of the course is maintained on file at the sponsoring institution and can be made available to NRC upon request;

- Evidence that the sponsoring institution has examined the student's knowledge of the course content is maintained on file at the institution and can be made available to NRC upon request. The evidence of the student's overall competency in the course material should include a final grade or percentile; and

- A permanent record that the student successfully completed the course is kept at the institution.

The AU must demonstrate training and experience with the type and quantity of material that is to be used at the pharmacy. For example, someone with training and experience only with microcurie quantities of unsealed radioactive material may not be qualified to use or supervise the use of higher activity sealed radioactive sources for instrument calibration. Applicants should pay particular attention to the type of radiation involved. For example, someone experienced with gamma emitters may not have appropriate experience for high-energy beta emitters.

Note that for applicants that produce radioactive material using an accelerator, the individual who handles byproduct materials during the maintenance and repair of an accelerator or other related equipment should also be considered an AU. However, training and experience documentation for these individuals should be submitted with the license application for radionuclide production as specified in NUREG-1556, Vol. 21, "Consolidated Guidance About Materials Licenses: Program-Specific Guidance About Possession Licenses for Production of Radioactive Materials Using an Accelerator."

Response from Applicant: For each proposed AU:

- Name of each proposed AU;

AND

- Types, quantities, and proposed uses of licensed material;

AND

- A copy of the license (NRC or Agreement State) on which the individual was specifically named as an AU for the types, quantities, and proposed uses of licensed materials;

OR

- A copy of the permit maintained by a licensee of broad scope that identifies the individual as an AU for the types, quantities, and proposed uses of licensed materials;

OR

- Description of the training and experience demonstrating that the proposed AU is qualified by training and experience to use the requested licensed materials. The applicant may find it convenient to describe this training and experience using a format similar to Tables G-1 and G-2 in Appendix G.

8.8 ITEM 8: TRAINING FOR INDIVIDUALS WORKING IN OR FREQUENTING RESTRICTED AREAS

8.8.1 OCCUPATIONALLY EXPOSED WORKERS AND ANCILLARY PERSONNEL

Regulations: 10 CFR 19.12, 10 CFR 20.1101(a), 10 CFR 30.33(a)(3).

Criteria: Individuals working with licensed material must receive radiation safety training commensurate with their assigned duties and specific to the licensee's Radiation Safety Program. In addition, those individuals who, in the course of employment, are likely to receive in a year a dose in excess of 100 mrem (1 mSv) must be instructed according to 10 CFR 19.12.

Discussion: Under 10 CFR 20.1101(a), each licensee is required to develop, document, and implement a Radiation Protection Program commensurate with the scope and extent of licensed activities and sufficient to ensure compliance with 10 CFR Part 20. Each individual working with radioactive material must be trained in the radiation safety procedures applicable to his/her job before beginning work with licensed materials. Licensees should not assume that safety instruction has been adequately covered by prior employment or training. Practical, site-specific training should be provided for all individuals prior to beginning work with, or in the vicinity of, licensed material. Training should also be performed whenever there is a significant change in duties, procedures, regulations, or terms of the license. Each individual should also receive periodic refresher training at least annually to ensure that all staff remain adequately trained.

Additional training is required if an individual is likely to receive a dose in excess of 1 mSv (100 mrem) in a year. ANPs and others involved in the preparation of radiopharmaceuticals are most likely to receive doses in excess of 1 mSv (100 mrem) in a year; however, potential radiation doses received by all employees must also be evaluated. The evaluation must include consideration of assigned activities during both normal and abnormal situations involving exposure to radiation and/or radioactive material that can reasonably be expected to occur during licensed activities.

If individuals making deliveries of radioactive material at the licensee's facility are likely to receive a dose in excess of 1 mSv (100 mrem) in a year from the licensee's activities, the licensee is responsible for ensuring that the person has received the training specified in 10 CFR Part 19, regardless of whether that person is an employee of the licensee. If the training has been provided by someone else (such as the shipper or another licensee), the licensee does not have to provide training except for instruction in site-specific radiation hazards. This issue is discussed in NRC Generic Letter 95-09, "Monitoring and Training of Shippers and Carriers of Radioactive Materials," dated November 3, 1995.

Training may be in the form of lecture, demonstrations, videotape, or self-study, and should emphasize practical subjects important to the safe use of licensed material. A method for asking questions should be provided for individuals receiving instructions and training. The licensee should determine whether the training succeeded in conveying the desired information and adjust the training program as necessary. The person conducting the training should be a qualified individual (e.g., RSO, ANP, AU, or radiation safety professional familiar with the licensee's program).

Licensee personnel who work in the vicinity of, but do not handle radioactive materials (ancillary staff), are not required to have radiation safety training as long as they are not likely to receive 1 mSv (100 mrem) in a year; however, to minimize potential radiation exposure when ancillary staff are working in the vicinity of radioactive material, it is prudent for them to work under the supervision and in the physical presence of an ANP/AU or to be provided some basic radiation safety training. Such ancillary staff should be informed of the nature and location of the radioactive material and the meaning of the radiation symbol, and should be instructed not to handle radioactive material and to keep away from it as much as their work permits.

Some ancillary staff, although not likely to receive doses over 1 mSv (100 mrem), should receive training to ensure adequate security and control of licensed material. Licensees may provide these individuals with training commensurate with their assignments in the vicinity of the radioactive material to ensure the control and security of the material.

The guidance in Appendix N may be used by the applicant to develop a training program.

Response from Applicant: State: "We have developed and will implement and maintain written procedures for a training program for each group of workers, including: topics covered; qualifications of the instructors; method of training; method for assessing the success of the training; and the frequency of training and refresher training."

References: For hard copies of NRC Generic Letter 95-09, "Monitoring and Training of Shippers and Carriers of Radioactive Materials," dated November 3, 1995, see the Notice of Availability (on the inside front cover of this report).

8.8.2 PERSONNEL INVOLVED IN HAZARDOUS MATERIALS PACKAGE PREPARATION AND TRANSPORT

Regulation: 49 CFR 172.700, 49 CFR 172.702, 49 CFR 172.704.

Criteria: Applicants must train personnel involved in the preparation and transport of hazardous material packages in the applicable DOT regulations.

Discussion: Licensees who prepare packages of radioactive materials or who transport their own packages must provide training to their employees who perform those functions. The training must include:

- General awareness and familiarization training designed to provide familiarity with DOT requirements, and the ability of the employee to recognize and identify hazardous materials;

- Function-specific training concerning the DOT requirements that are specifically applicable to the functions the employee performs (e.g., if the employee's duties require affixing DOT radioactive labels to packages, the employee must receive training in DOT's regulations governing package labeling);

- Safety training concerning emergency response information, discussed above; measures to protect the employee and other employees from the hazards associated with the hazardous materials to which they may be exposed in the workplace; and methods of avoiding accidents, such as the proper procedures for handling packages containing hazardous materials; and

- Security awareness training: training regarding awareness of security risks associated with hazardous materials transportation and methods designed to enhance transportation security.

The training must be provided initially, and every 3 years thereafter. Records of training must be maintained.

Note: The licensee is not responsible for providing DOT-required hazardous materials training to common carriers to whom the pharmacy offers radioactive materials packages for transport.

Response from Applicant: Submit the following statement: "We have developed and will implement and maintain written procedures for training personnel involved in hazardous materials package preparation and transport that meet the requirements in 49 CFR 172.700, 49 CFR 172.702, and 49 CFR 172.704, as applicable."

8.8.3 INSTRUCTION FOR SUPERVISED INDIVIDUALS PREPARING RADIOPHARMACEUTICALS

Regulations: 10 CFR 32.72(b)(1), 10 CFR 35.27(b).

Criteria: Individuals who prepare byproduct material for medical use under the supervision of an authorized nuclear pharmacist must be instructed in the preparation of byproduct material for medical use, the principles of radiation safety, and the licensee's procedures for the use of byproduct material; must follow the instructions given; and must have their work, and records kept to reflect their work, periodically reviewed by the supervising ANP.

Discussion: The applicant must instruct supervised individuals in the preparation of byproduct material for medical use and require those individuals to follow their instructions, the written Radiation Protection Program, license conditions, and NRC regulations. The supervising ANP must review the work of supervised individuals in the preparation of byproduct material for medical use and the records kept to reflect that work. If an ANP is always physically present when radioactive drugs are prepared, supervision may be fulfilled by the day-to-day instruction and review of the supervised individual by the ANP.

An ANP is considered to be supervising the use of radioactive materials when directing personnel in the conduct of operations involving licensed materials. The ANP need not be present at all times during the use of such materials; however, the supervising ANP is responsible for ensuring that personnel under supervision have been properly trained and instructed. The supervising ANP is therefore responsible for the supervision of operations involving the use of radioactive materials, whether or not he or she is present.

The NRC regulations do not relieve the licensee from complying with applicable Department of Health and Human Services (Food and Drug Administration), other Federal, and State requirements governing radioactive drugs. From an NRC perspective, if the supervision requirements are met, it is permissible for the licensee to allow the supervised individual to prepare radiopharmaceuticals without the presence of the ANP; however, some States require that a pharmacist be physically present during the preparation and dispensing of pharmaceuticals, including radioactive drugs. It is the licensee's responsibility to ensure that its practices comply with any additional State requirements concerning this issue.

Response from Applicant: No response from the applicant is necessary. Supervision will be reviewed during inspection.

8.9 ITEM 9: FACILITIES AND EQUIPMENT

8.9.1 FACILITIES AND EQUIPMENT FOR RADIOPHARMACIES

Regulations: 10 CFR 32.72(a)(2), 10 CFR 30.33(a)(2), 10 CFR 20.1406, 10 CFR 20.1101(b), 10 CFR 30.35 (g).

Criteria: A radiopharmacy must demonstrate that it is a pharmacy. Facilities and equipment must be adequate to protect health and minimize danger to life or property, minimize the likelihood of contamination, and keep exposures to workers and the public ALARA.

Discussion: An applicant must demonstrate that it is a pharmacy by submitting evidence of at least one of the following:

- Licensure as a pharmacy by a State Board of Pharmacy, or

- Operation as a nuclear pharmacy within a Federal medical institution.

If the registration or license has not been issued by the State Board of Pharmacy at the time of application, the applicant may provide it at a later date, but prior to license issuance from NRC.

Applicants must provide NRC with documentation demonstrating that their facilities and equipment provide sufficient engineering controls and barriers to protect the health and safety of the public and their employees. The facilities and equipment must also keep exposures to radiation and radioactive materials ALARA and minimize the risks from the uses of the types and quantities of radioactive materials. The applicant should provide clear delineations between its restricted and unrestricted areas through the use of barriers, postings, and worker instructions.

Applicants may delay completing facilities and acquiring equipment until after the application review is completed, in case changes are required as a result of the application review. This also ensures the adequacy of the facilities and equipment before the applicant makes a significant financial commitment. In all cases, the applicant cannot possess or use licensed material until after the facilities are approved, equipment is procured, and the license is issued.

It is important to note that applicants who plan to amend their license to add the use and distribution of high-energy gamma-/photon-emitting radionuclides, such as PET radionuclides, to their operations should ensure their facilities and equipment are adequate to handle the higher radiation levels. Most likely, applicants will need to add and/or replace shielding, modify ventilation and air filtration systems, and possibly modify the facility's design to accommodate the higher energy radionuclides.

Applicants are reminded that records important to decommissioning are required to be maintained in an identifiable location. For further information, see the section entitled, "Financial Assurance and Recordkeeping for Decommissioning."

Response from Applicant: Applicants must provide: Copies of their registration or license from a State Board of Pharmacy as a pharmacy, or evidence that they are operating as a nuclear pharmacy within a Federal medical institution;

Note: If the applicant's particular activities are not recognized as the practice of commercial radiopharmacy, the applicant must submit evidence that it is registered or licensed with the State or FDA as a drug manufacturer. Refer to NUREG-1556, Volume 12, for guidance on drug manufacturer requirements.

AND

- Description of the facilities and equipment to be made available at each location where radioactive material will be used. A diagram should be submitted showing the applicant's entire facility and identifying activities conducted in all contiguous areas surrounding the facility (see Figure 8.3). Diagrams should be drawn to a specified scale, or dimensions should be indicated.

Include the following information:

- Descriptions of the area(s) assigned for the receipt, storage, preparation, and measurement of radioactive materials and the location(s) for radioactive waste storage;

- Sufficient detail in the diagram to indicate locations of shielding, the proximity of radiation sources to unrestricted areas, and other items related to radiation safety;

- A general description of the ventilation system, including representative equipment such as glove boxes or fume hoods. Pertinent airflow rates, differential pressures, filtration equipment, and monitoring systems should be described in terms of the minimum performance to be achieved. Confirm that such systems will be employed for the use or storage of radioactive materials likely to become airborne, such as compounding radioiodine capsules and dispensing radioiodine solutions;

- Verification that ventilation systems ensure that effluents are ALARA, are within the dose limits of 10 CFR 20.1301, and are within the ALARA constraints for air emissions established under 10 CFR 20.1101(d); and

- Mark drawings and diagrams that provide exact location of materials or depict specific locations of safety or security equipment as, "Security-Related Information – Withhold Under 10 CFR 2.390."

SECURITY-RELATED INFORMATION - WITHHOLD UNDER 10 CFR 2.390*

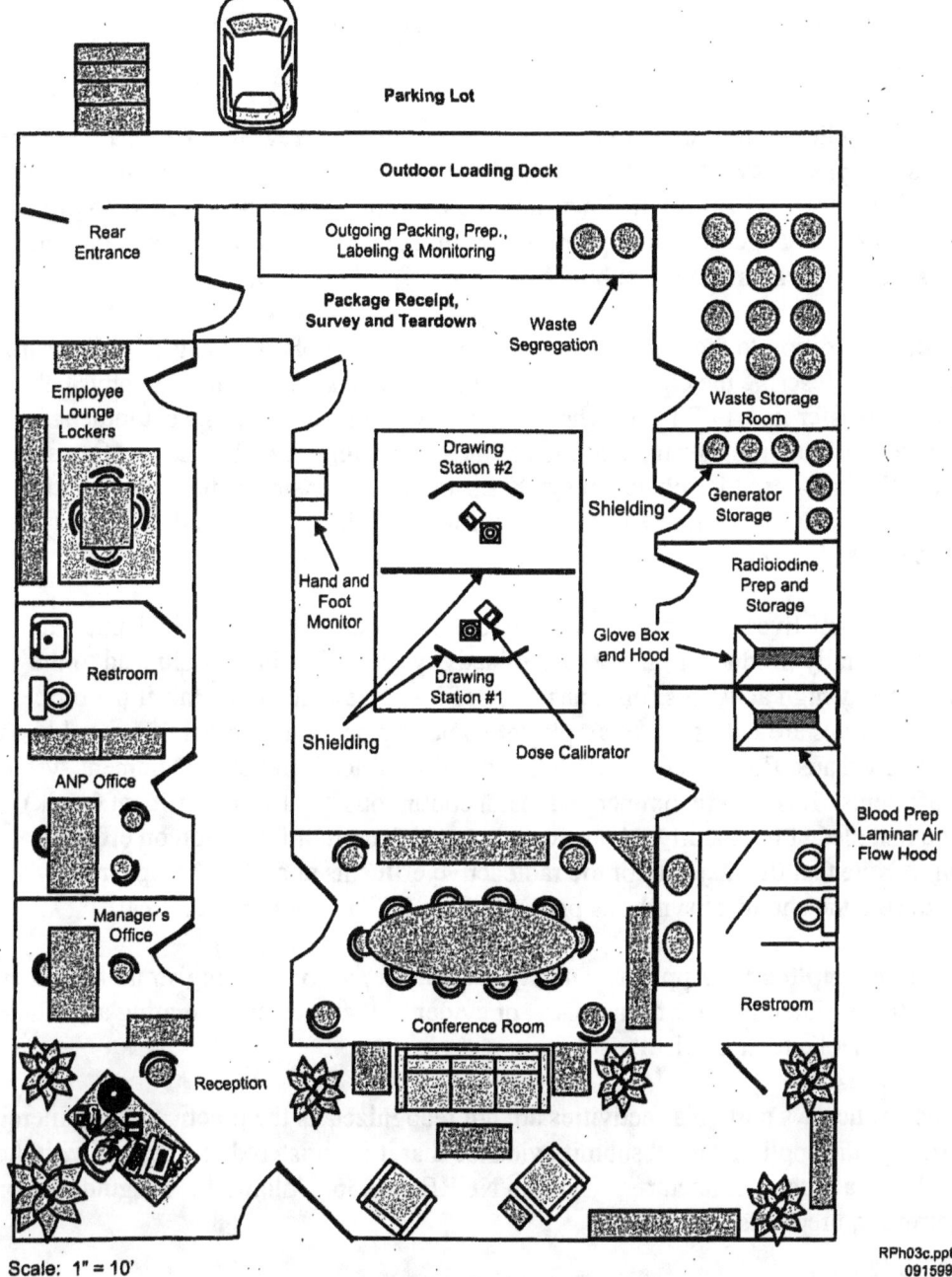

Scale: 1" = 10'

RPh03c.ppt
091599

SECURITY-RELATED INFORMATION - WITHHOLD UNDER 10 CFR 2.390*

* For the purposes of this NUREG, this diagram is marked appropriately for an application. This particular diagram does not contain real security-related information.

Figure 8.3 Typical Radiopharmacy Diagram.

8.9.2 FACILITIES AND EQUIPMENT FOR PET RADIOPHARMACIES

Regulations: 10 CFR 32.72(a)(2), 10 CFR 30.33(a)(2), 10 CFR 20.1406, 10 CFR 20.1101(b), 10 CFR 30.35 (g).

Criteria: PET radiopharmacies must demonstrate that they are registered with a State agency, are licensed as a pharmacy by the State's Board of Pharmacy, or operate as a nuclear pharmacy within a Federal medical institution. Facilities and equipment must be adequate to protect health and minimize danger to life or property, minimize the likelihood of contamination, and keep exposures to workers and the public ALARA.

Discussion: In addition to the information required for a radiopharmacy, PET radiopharmacy applicants should describe the equipment and/or method and shielding used to physically transfer (e.g., transfer lines) PET radiochemicals to the chemical synthesis equipment for radiopharmaceutical manufacturing and then to the dispensing area. The description should also include shielding used for chemical synthesis and/or dispensing radiopharmaceuticals. Also, the type of remote handling equipment used for handling the PET radionuclides and drugs should be described.

Due to the short half-lives of positron-emitting radionuclides, commercial PET nuclear pharmacies generally produce high amounts of activity (curies), which could lead to the potential for fairly high activities (millicuries) of effluents released in the air if the proper engineering controls are not used. Examples of some engineering controls that should be used would include exhaust filtration (e.g., HEPA and carbon filters) and/or containment systems for decay of effluents. It is also recommended that a continuous "real-time" effluent (stack) monitor be installed at the facility. Appendix R provides more information on effluent monitoring. Note that the majority of the radioactive effluents at a PET radiopharmacy are produced during the chemical synthesis process of the PET radiopharmaceutical.

Response from Applicant: Applicants must provide: Copies of their registration or license as a pharmacy from a State Board of Pharmacy, or evidence that they are operating as a nuclear pharmacy within a Federal medical institution;

Note: If the applicant's particular activities are not recognized as the practice of commercial radiopharmacy, the applicant must submit evidence that it is registered or licensed with the State or FDA as a drug manufacturer. Refer to NUREG-1556, Volume 12, for guidance on drug manufacturer requirements.

AND

- Description of the facilities and equipment to be made available at each location where radioactive material will be used, which includes the method and shielding used to physically transfer licensed material (e.g., transfer lines) to the different processes (e.g., chemical synthesis, dispensing). A diagram should be submitted that shows the applicant's entire facility and identifies activities conducted in all contiguous areas surrounding the facility (see

Figure 8.4). Diagrams should be drawn to a specified scale, or dimensions should be indicated.

Include the following information:

- Descriptions of the area(s) assigned for the production or receipt, storage, preparation, measurement, and distribution of radioactive materials and the location(s) for radioactive waste storage;

- Sufficient detail in the diagram to indicate locations of shielding and/or shielding equipment (e.g., hot cells for positron-emitting radionuclides), the proximity of radiation sources to unrestricted areas, and other items related to radiation safety, such as remote handling equipment and area monitors;

- A general description of the ventilation system, including representative equipment such as glove boxes or fume hoods. Pertinent airflow rates, differential pressures, filtration equipment, and monitoring systems should be described in terms of the minimum performance to be achieved. Confirm that such systems will be employed for the production, use, or storage of radioactive materials; and

- Verification that ventilation systems ensure that effluents are ALARA, are within the dose limits of 10 CFR 20.1301, and are within the ALARA constraints for air emissions established under 10 CFR 20.1101(d).

SECURITY-RELATED INFORMATION - WITHHOLD UNDER 10 CFR 2.390*

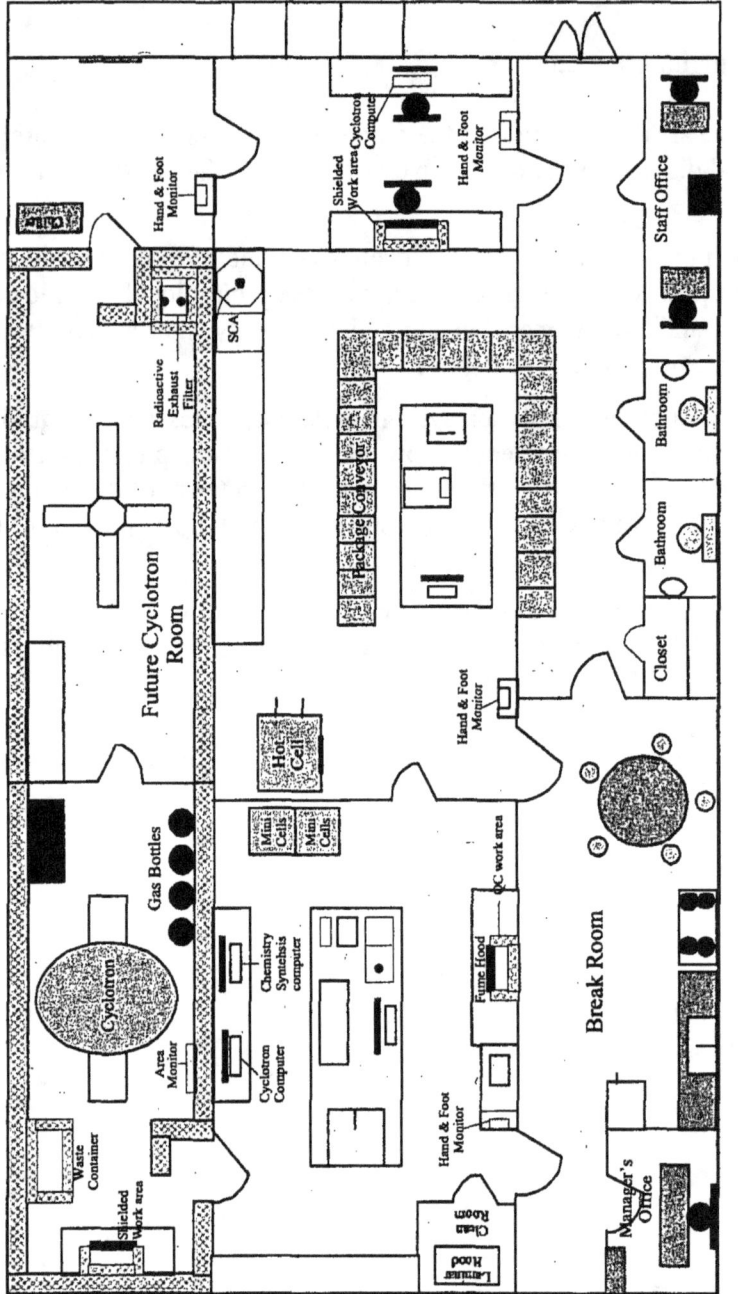

SECURITY-RELATED INFORMATION - WITHHOLD UNDER 10 CFR 2.390*

* For the purposes of this NUREG, this diagram is marked appropriately for an application. This particular diagram does not contain real security-related information.

Figure 8.4 Typical PET Radiopharmacy Diagram

8.10 ITEM 10: RADIATION SAFETY PROGRAM

8.10.1 AUDIT PROGRAM

Regulations: 10 CFR 20.1101, 10 CFR 20.2102.

Criteria: Licensees must review the content and implementation of their Radiation Protection Programs annually to ensure the following:

- Compliance with NRC and DOT regulations (as applicable) and the terms and conditions of the license,

- Occupational doses and doses to members of the public are ALARA (10 CFR 20.1101), and

- Records of audits and other reviews of program content are maintained for 3 years.

Discussion: Appendix I contains a suggested audit program that is specific to commercial radiopharmacies and is acceptable to NRC. All areas indicated in Appendix I may not be applicable to every licensee, and all items may not need to be addressed during each audit. For example, licensees do not need to address areas that do not apply to their activities, and activities which have not occurred since the last audit need not be reviewed at the next audit.

Currently, NRC's emphasis during inspections is to perform actual observations of work in progress. As a part of their audit programs, applicants should consider performing unannounced audits of the radiopharmacy to observe whether radiation safety procedures are being followed, etc.

It is essential that once identified, problems be corrected comprehensively and in a timely manner; Information Notice (IN) 96-28, "Suggested Guidance Relating to Development and Implementation of Corrective Action," provides guidance on this subject. NRC will review the licensee's audit results and determine if corrective actions are thorough, timely, and sufficient to prevent recurrence. If violations are identified by the licensee and these steps are taken, NRC can exercise discretion and will normally elect not to cite a violation. NRC's goal is to encourage prompt identification and prompt comprehensive correction of violations and deficiencies. For additional information on NRC's use of discretion on issuing violations, refer to the current version of NRC's Enforcement Policy.

Licensees must maintain records of audits and other reviews of program content and implementation for three years from the date of the record. NRC has found audit records that contain the following information to be acceptable: date of audit, name of person(s) who conducted the audit, persons contacted by the auditor(s), areas audited, audit findings, corrective actions, and follow-up.

Response from Applicant: No response is required. The licensee's program for auditing its Radiation Safety Program will be reviewed during inspection.

References: The current version of NRC's Enforcement Policy is included on NRC's Web site at http://www.nrc.gov/about-nrc/regulatory/enforcement.html. The INs are available in the "Electronic Reading Room" on NRC's Internet Page at http://www.nrc.gov/reading-rm/doc-collections/gen-comm/info-notices.

8.10.2 RADIATION MONITORING INSTRUMENTS

Regulations: 10 CFR 20.1501, 10 CFR 32.72(c), 10 CFR 30.33(a)(2), 10 CFR 20.2103(a).

Criteria: Licensees must possess radiation monitoring instruments to evaluate possible radiation hazards that may be present. Instruments used for quantitative radiation measurements must be calibrated periodically for the radiation measured.

Discussion: Licensees must possess calibrated radiation detection/measurement instruments to perform, as necessary, the following:

- Package surveys,
- Personnel and facility contamination measurements,
- Sealed-source leak tests,
- Air sampling measurements,
- Bioassay measurements,
- Effluent release measurements, and
- Dose-rate surveys.

For the purposes of this document, radiation monitoring instruments are defined as any device used to measure the radiological conditions at a licensed facility. Some of the instruments that may be used to perform the above functions include:

- Portable or stationary count rate meters,
- Portable or stationary dose rate or exposure rate meters,
- Area Monitors,
- Single or multichannel analyzers,
- Liquid Scintillation Counters (LSC),
- Gamma counters,
- Proportional counters,
- Solid-state detectors, and
- Hand and foot contamination monitors.

The choice of instrument should be appropriate for the type of radiation to be measured and for the type of measurement to be taken (count rate, dose rate, etc.). Radiopharmacies typically use a broad energy range of gamma and beta radiation emitters and need to use radiation detectors appropriate for those energies. Figure 8.5 illustrates some common survey instruments used for contamination surveys. Applicants should discuss the types of instruments to be used for each type of survey to be performed and the availability of a sufficient quantity of these instruments at their facility.

Hand and Foot Monitor

Contamination Detector

Beta/Gamma Probes

Survey Meter and Attached Beta/Gamma Probe

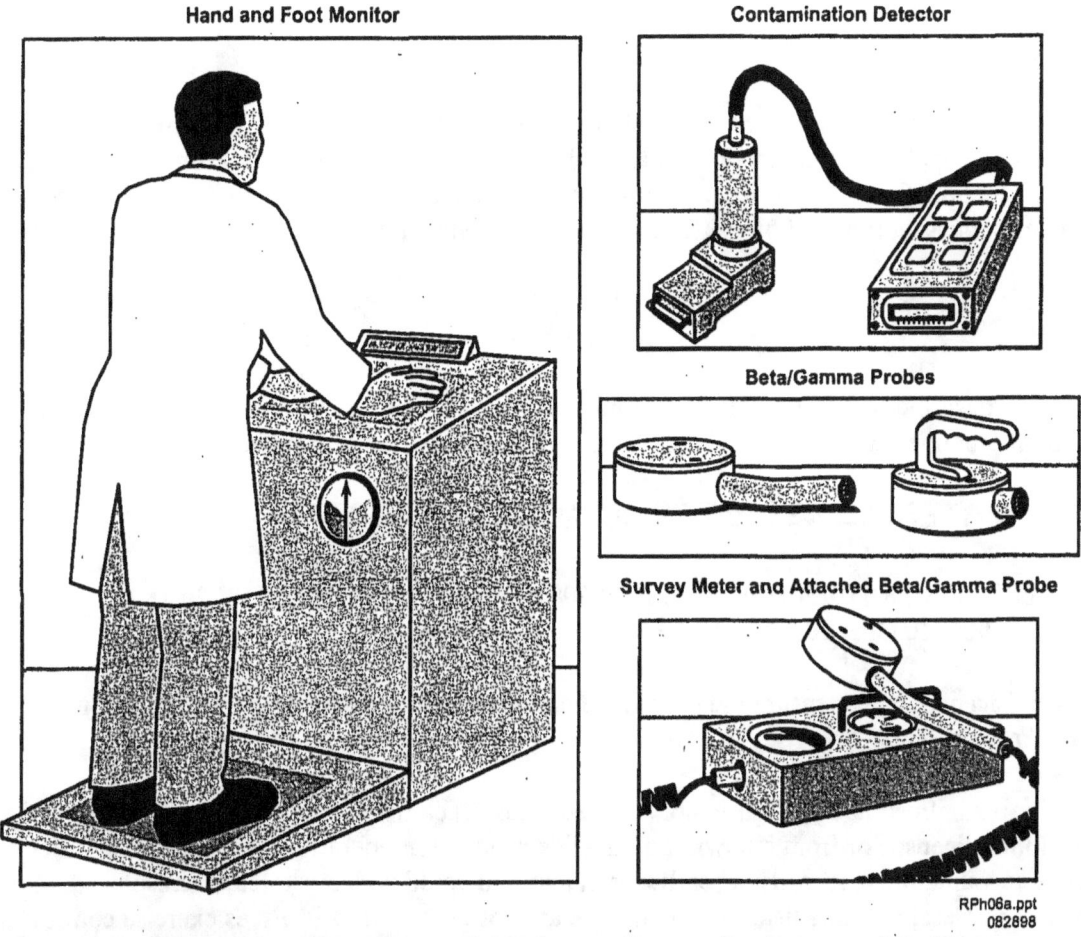

RPh06a.ppt
082898

Figure 8.5 Examples of Portable and Stationary Survey Instruments Used by Radiopharmacies.

Instrument calibrations may be performed by the pharmacy or by another person specifically authorized by NRC, an Agreement State, or a licensing State to perform that function. If the pharmacy utilizes the services of another person for instrument calibration, the pharmacy should ensure that person has been authorized by either NRC, an Agreement State, or a licensing State to perform that activity. Appendix J provides information about general instrument specifications and model calibration procedures.

Response from Applicant: Do one of the following:

- State: "We will use equipment that meets the general radiation monitoring instrument specifications and implement the model survey meter calibration program published in Appendix J to NUREG-1556, Vol. 13, Rev. 1, 'Program-Specific Guidance About Commercial Radiopharmacy Licenses' ";

OR

- State: "We will use equipment that meets the general radiation monitoring instrument specifications published in Appendix J to NUREG-1556, Vol. 13, Rev. 1, 'Program-Specific Guidance About Commercial Radiopharmacy Licenses,' and instruments will be calibrated by other persons authorized by NRC, an Agreement State, or a licensing State to perform that service";

OR

- Provide a description of alternative minimum equipment to be used for radiation monitoring and/or alternative procedures for the calibration of radiation monitoring equipment.

8.10.3   MATERIAL RECEIPT AND ACCOUNTABILITY

Regulations: 10 CFR 20.1501(a), 10 CFR 20.2001, 10 CFR 20.1801, 10 CFR 20.1802, 10 CFR 20.1906, 10 CFR 20.2201, 10 CFR 30.41, 10 CFR 30.51.

Criteria: Licensees must ensure the security and accountability of licensed material and must open packages safely.

Discussion: As illustrated in Figure 8.6, licensees must track licensed materials from receipt (from another licensee or from its own radionuclide production operations) to disposal in order to ensure accountability; identify when licensed material could be lost, stolen, or misplaced; and ensure that possession limits listed on the license are not exceeded. Licensees exercise control over licensed material accountability by including the following items (as applicable) in their Radiation Protection Program:

- Physical inventories of sealed sources at intervals not to exceed 6 months,
- Ordering and receiving licensed material,
- Package opening,
- Maintaining material inventory within license possession limits,
- Transfer of material, including distribution, and
- Disposal of material.

Licensees are required to develop, implement, and maintain written procedures for safely opening packages in accordance with 10 CFR 20.1906. Some packages may require special procedures that take into consideration the type, quantity, or half-life of the nuclide being delivered.

A model procedure for safely opening packages containing licensed materials is included in Appendix P.

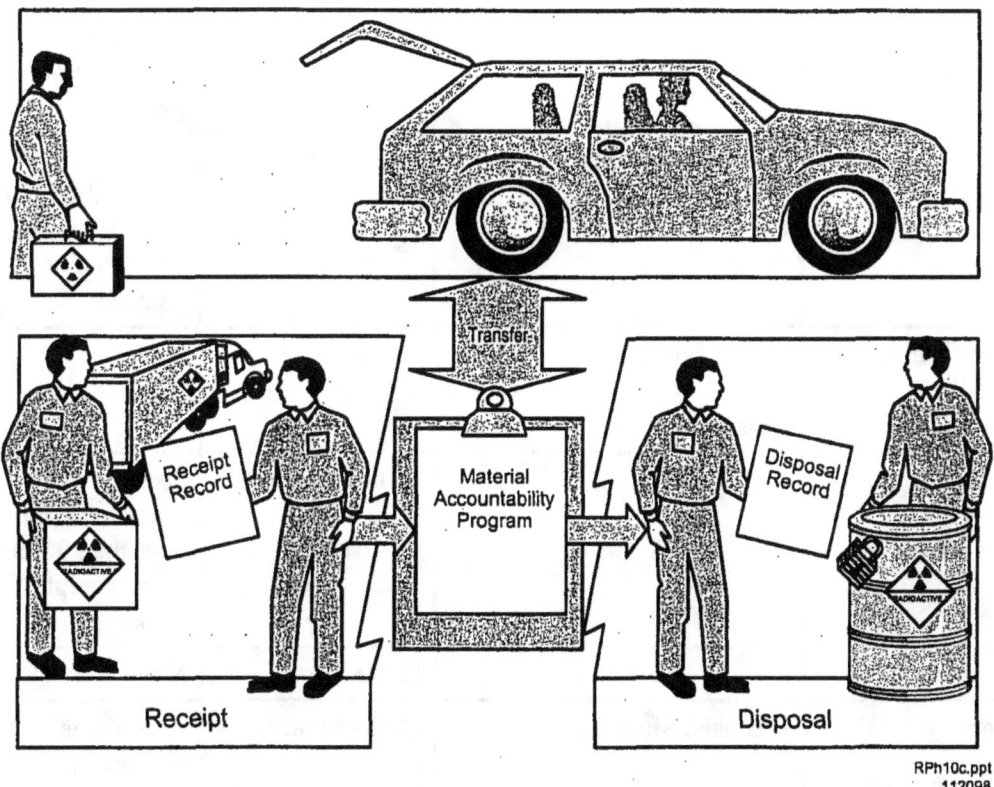

RPh10c.ppt
112098

Figure 8.6 Accountability. *Licensees must maintain records of receipt (from another licensee or from its own production operations), transfer, and disposal of licensed material.*

NRC regulations in 10 CFR 20.1906(b) and (c) state the requirements for monitoring packages containing licensed material. These requirements are described in Table 8.1, below.

Under 10 CFR 20.1906(d), the licensee is required to immediately notify the final delivery carrier and NRC Operations Center when removable radioactive surface contamination exceeds the limit of 22 disintegrations per minute per square centimeter (dpm/cm^2) averaged over 300 cm^2; or external radiation levels exceed 2.0 mSv/hr (200 mrem/hr) at the surface.

Licensees must secure and control licensed material and should have a means of promptly detecting losses of licensed material. NRC regulations in 10 CFR 20.1801 and 20.1802 require licensees to secure radioactive materials from unauthorized removal or access while in storage and to control and maintain constant surveillance over licensed material that is not in storage.

Table 8.1 Package Monitoring Requirements

Package	Contents	Survey Type	Survey Time*
Labeled (White I, Yellow II, Yellow III)	Gas or Special Form Greater Than Type A	Radiation Level	As soon as practicable, but not later than 3 hours after receipt of package
Labeled (White I, Yellow II, Yellow III)	Not Gas or Special Form Greater Than Type A	Contamination and Radiation Level	As soon as practicable, but not later than 3 hours after receipt of package
Labeled (White I, Yellow II, Yellow III)	Gas or Special Form Less Than Type A	None	None
Labeled (White I, Yellow II, Yellow III)	Not Gas or Special Form Less Than Type A	Contamination	As soon as practicable, but not later than 3 hours after receipt of package
Not Labeled	Licensed Material	None	None
Damaged	Licensed Material	Contamination and Radiation Level	As soon as practicable, but not later than 3 hours after receipt of package

* Assumes packages are received during normal working hours. If packages are received outside of normal working hours, the licensee has three hours after the beginning of the next work day to perform the required surveys.

Licenses will normally contain specific conditions requiring the licensee to perform inventories and leak tests of sealed sources every six months (see sample license in Appendix E). Since the leak tests require an individual to locate and work with the sealed source, records of leak tests may be used as part of an inventory and accountability program. Sources in storage that are used infrequently may not require leak testing; however, the inventory must still be performed at the specified interval.

With regard to unsealed licensed material, licensees use various methods (e.g., computer programs, manual ledgers, log books) to account for receipt, use, transfer, disposal, and radioactive decay. These methods help to ensure that possession limits are not exceeded.

Table 8.2 lists the types and retention times for the records of receipt, use, transfer, and disposal (as waste) of all licensed material the applicant must maintain. Other records, such as transfer records, could be linked to radioactive material inventory records.

Table 8.2 Record Maintenance

Type of Record	How Long Record Must be Maintained
Receipt	For as long as the material is possessed until 3 years after transfer or disposal
Transfer	For 3 years after transfer
Disposal	Until NRC terminates the license
Important to decommissioning	Until the site is released for unrestricted use

Material accountability records typically contain the following information:

- Radionuclide and activity (in units of becquerels or curies), and date of measurement of byproduct material;

- For each sealed source, manufacturer, model number, location and, if needed for identification, serial number and as appropriate, manufacturer and model number of device containing the sealed source;

- Date of the transfer and name and license number of the recipient, and description of the radioactive material (e.g., radionuclide, activity, manufacturer's name and model number, serial number); and

- For licensed materials disposed of as waste, the radionuclide, activity, date of disposal, and method of disposal (decay, sewer, etc.).

See the section on "Waste Disposal" for additional information.

Information about locations where licensed material is used or stored is among the records important to decommissioning and required by 10 CFR 30.35(g). See also the section on "Financial Assurance and Recordkeeping for Decommissioning."

Response from Applicant: Provide the following statements:

- "We have developed, and will implement and maintain, written procedures for safely opening packages that meet the requirements in 10 CFR 20.1906";

AND

- "We will conduct physical inventories of sealed sources of licensed material at intervals not to exceed 6 months";

AND

- "We have developed, and will implement and maintain written procedures for licensed material accountability and control to ensure that:

  — license possession limits are not exceeded,

  — licensed material in storage is secured from unauthorized access or removal,

  — licensed material not in storage is maintained under constant surveillance and control, and

  — records of receipt, either from the licensee's own production operations or from another licensee transfer, and disposal of licensed material, are maintained."

8.10.4 OCCUPATIONAL DOSE

Regulations: 10 CFR 20.1501, 10 CFR 20.1502, 10 CFR 20.1201, 10 CFR 20.1202, 10 CFR 20.1203, 10 CFR 20.1204, 10 CFR 20.1207, 10 CFR 20.1208, 10 CFR 20.2106, 10 CFR Part 20 Appendix B.

Criteria: Each licensee shall evaluate the potential occupational exposures of all workers and monitor occupational exposure to radiation when required.

Discussion: The licensee should perform an evaluation of the dose the individual is likely to receive prior to allowing the individual to receive the dose (prospective evaluation). When performing the prospective evaluation, only a dose that could be received at the facilities of the applicant or licensee performing the evaluation needs to be considered. These estimates can be based on any combination of work location radiation monitoring, survey results, monitoring results of individuals in similar work situations, or other estimates to produce a "best estimate" of the actual dose received. For individuals who have received doses at other facilities in the current year, the previous dose need not be considered in the prospective evaluation if monitoring was not required at the other facilities. This evaluation need not be made for every individual; evaluations can be made for employees with similar job functions or work areas. Further guidance on evaluating the need to provide monitoring is provided in Regulatory Guide 8.34, "Monitoring Criteria and Methods to Calculate Occupational Doses," dated July 1992.

If the prospective evaluation shows that an individual's dose is not likely to exceed 10% of any applicable regulatory limit, the individual is not required to be monitored for radiation exposure and there are no recordkeeping or reporting requirements for doses received by that individual. If the prospective dose evaluation shows that the individual is likely to exceed 10% of an applicable limit, monitoring is required.

Licensees shall monitor worker exposures for:

Adults who are likely to receive an annual dose in excess of any of the following:

- 5 mSv (0.5 rem) deep-dose equivalent,

- 15 mSv (1.5 rems) eye dose equivalent,

- 50 mSv (5 rems) shallow-dose equivalent to the skin, and

- 50 mSv (5 rems) shallow-dose equivalent to any extremity.

Minors who are likely to receive an annual dose in excess of any of the following:

- 1.0 mSv (0.1 rem) deep-dose equivalent,

- 1.5 mSv (0.15 rem) eye dose equivalent,

- 5 mSv (0.5 rem) shallow-dose equivalent to the skin, and

- 5 mSv (0.5 rem) shallow-dose equivalent to any extremity.

Declared pregnant women who are likely to receive an annual dose from occupational exposures in excess of 1.0 mSv (0.1 rem) deep-dose equivalent, although the dose limit applies to the entire gestation period.

Internal exposure monitoring is required for:

- Adults likely to receive in 1 year an intake in excess of 10% of the applicable ALIs for ingestion and inhalation, and

- Minors and declared pregnant women likely to receive in 1 year a committed effective dose equivalent in excess of 1.0 mSv (0.1 rem).

If an individual is likely to receive in 1 year a dose greater than 10% of any applicable limit (see Figure 8.7 for annual dose limits for adults), monitoring for occupational exposure is required. Radiopharmacy technologists and ANPs are generally likely to receive 10% of the limits for occupational dose. Most radiopharmacies provide these employees with whole body and extremity monitors.

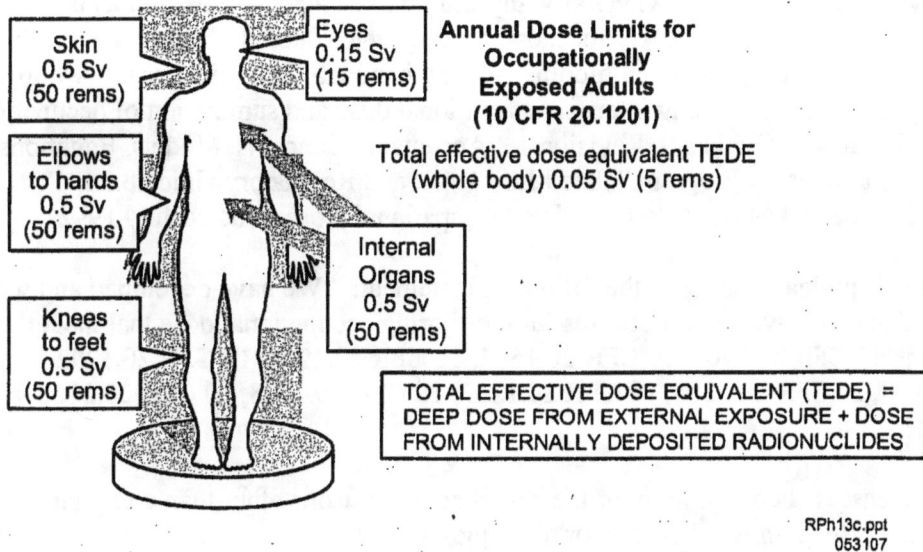

Figure 8.7 Annual Dose Limits for Occupationally Exposed Adults.

Most licensees use either film badges or thermoluminescent dosimeters (TLDs), or optically-stimulated luminescence (OSL) dosimeters that are supplied by a National Voluntary Laboratory Accreditation Program (NVLAP)-approved processor to monitor for external exposure. The exchange frequency for film badges is generally monthly due to technical concerns about film fading. The exchange frequency for TLDs and OSLs are generally quarterly. Applicants should verify that the processor is NVLAP-approved. Consult the NVLAP-approved processor for its recommendations for exchange frequency and proper use. If monitoring is required, then the licensee must maintain records of the monitoring regardless of the actual dose received. Also, when working with high-energy gamma-/photon-emitting radionuclides such as positron-emitting radionuclides, it is recommended that a pocket/audible dosimeter be worn by ANPs, AUs, and radiopharmacy technologists, in addition to their personal dosimeter(s).

The types and quantities of radioactive material used at most commercial radiopharmacies provide a reasonable possibility for an internal intake by ANPs and radiopharmacy technologists. Uses such as preparing radioiodine capsules from liquid solutions, and opening and dispensing from vials containing millicurie quantities of radioiodine and other isotopes, require particular caution. Precautionary measures for personnel to follow during iodine capsule preparation should involve the use of a fume hood and glove box or shoulder length gloves (see Appendix Q for additional guidance on precautionary measures). To monitor internal exposure from such operations, most pharmacies institute a routine bioassay program to periodically monitor these workers.

A program for performing thyroid uptake bioassay measurements should include adequate equipment to perform bioassay measurements, as well as procedures for calibrating the equipment, including factors necessary to convert counts per minute into becquerel or microcurie units, and should address the technical problems commonly associated with performing thyroid bioassays (e.g., statistical accuracy, attenuation by neck tissue). Thyroid bioassay procedures should also specify the interval between bioassays, action levels, and the actions to be taken at those levels. Generally, thyroid uptake bioassay measurements at radiopharmacies are performed weekly for those workers who routinely handle radioiodine or are in the immediate vicinity when radioiodine is being handled. For guidance on developing bioassay programs and determining internal occupational dose and summation of occupational dose, refer to Regulatory Guide 8.9, Revision 1, "Acceptable Concepts, Models, Equations and Assumptions for a Bioassay Program," dated July 1993, and Regulatory Guide 8.34, "Monitoring Criteria and Methods to Calculate Occupational Doses," dated July 1992.

Response from Applicant: Submit the following statement: "We have developed and will implement and maintain written procedures for monitoring occupational dose that meet the requirements in 10 CFR 20.1501, 10 CFR 20.1502, 10 CFR 20.1201, 10 CFR 20.1202, 10 CFR 20.1203, 10 CFR 20.1204, 10 CFR 20.1207, 10 CFR 20.1208, 10 CFR 20.2106, as applicable."

Note: Some licensees choose to monitor their workers for reasons other than compliance with NRC requirements (e.g., in response to worker requests).

References: National Institute of Standards and Technology (NIST) Publication 810, "National Voluntary Laboratory Accreditation Program Directory," is published annually and is available electronically at http://ts.nist.gov/nvlap. NIST Publication 810 can be purchased from GPO, whose URL is http://www.gpo.gov. ANSI N322 may be ordered electronically at http://www.ansi.org. Go to http://www.nrc.gov/reading-rm/doc-collections/reg-guides for Regulatory Guide 8.7, Revision 1, "Instructions for Recording and Reporting Occupational Radiation Exposure Data," dated June 1992; Regulatory Guide 8.9, Revision 1, "Acceptable Concepts, Models, Equations and Assumptions for a Bioassay Program," dated July 1993; and Regulatory Guide 8.34, "Monitoring Criteria and Methods to Calculate Occupational Radiation Doses," dated July 1992.

8.10.5 PUBLIC DOSE

Regulations: 10 CFR 20.1003, 10 CFR 20.1101(d), 10 CFR 20.1301, 10 CFR 20.1302, 10 CFR 20.1801, 10 CFR 20.1802, 10 CFR 20.2107, 10 CFR 20.2203.

Criteria: Licensees must do the following:

- Ensure that licensed material will be used, transported, stored, and disposed of in such a way that members of the public will not receive more than 1 mSv (100 mrem) (TEDE) in one year from licensed activities;

- Ensure that air emissions of radioactive material to the environment will not result in exposures to individual members of the public in excess of 0.1 mSv (10 mrem) (TEDE) in one year from those emissions;

- Ensure that the dose in any unrestricted area will not exceed 0.02 mSv (2 mrem) in any one hour, from licensed operations; and

- Prevent unauthorized access, removal, or use of licensed material.

Discussion: "Member of the public" is defined in 10 CFR 20.1003 as "any individual except when that individual is receiving an occupational dose." "Public dose" is defined as "the dose received by a member of the public from exposure to radiation and/or radioactive material released by a licensee, or to any other source of radiation under the control of a licensee." Public dose excludes doses received from background radiation, from sanitary sewerage discharges from licensees, and from medical procedures. Whether the dose to an individual is an occupational dose or a public dose depends on the individual's assigned duties. It does not depend on the area (restricted, controlled, or unrestricted) the individual is in when the dose is received. For guidance about accepted methodologies for determining dose to members of the public, refer to Appendix K.

There are many possible internal dose pathways that contribute to the TEDE. The TEDE can, however, be broken down into three major dose pathway groups:

1. Airborne radioactive material,

2. Waterborne radioactive material, and

3. External radiation exposure.

The licensee should review these major pathways and decide which are applicable to its operations. The licensee must ensure that the TEDE from all exposure pathways arising from licensed activities does not exceed 1.0 mSv (100 mrem) to the maximally exposed member of the public. In addition, the licensee must control air emissions, such that the individual member of the public likely to receive the highest TEDE does not exceed the constraint level of 0.1 mSv (10 mrem) per year from those emissions. If exceeded, the licensee must report this, in accordance with 10 CFR 20.2203, and take prompt actions to ensure against recurrence.

Licensees should design a monitoring program to ensure compliance with 10 CFR 20.1101(d) and 10 CFR 20.1302(b). The extent and frequency of monitoring will depend upon each licensee's needs. For additional guidance regarding monitoring of effluents, refer to the section entitled, "Radiation Safety Program - Surveys."

During NRC inspections, licensees must be able to provide documentation demonstrating, by measurement or calculation, that the TEDE to the individual member of the public likely to receive the highest dose from the licensed operation does not exceed the annual limit and the dose constraint. See Appendix K for examples of methods to demonstrate compliance.

Response from Applicant: No response is required from the applicant in a license application, but records demonstrating compliance will be examined during inspection.

8.10.6   SAFE USE OF RADIONUCLIDES AND EMERGENCY PROCEDURES

Regulations: 10 CFR 20.1101, 10 CFR 20.1801, 10 CFR 20.1802, 10 CFR 20.2201, 10 CFR 20.2202, 10 CFR 20.2203, 10 CFR 30.34(g), 10 CFR 30.50, 10 CFR 19.11(a)(3).

Criteria: Licensees are required to do the following:

- Keep radiation doses to workers and members of the public ALARA,

- Ensure security of licensed material, and

- Make the required notifications of events to NRC.

Discussion: Licensees are responsible for the security and safe use of all licensed material from the time it arrives or is produced at their facility until its use, transfer, and/or disposal. Licensees should develop written procedures to ensure safe use of licensed material, and the procedures should also include operational and administrative guidelines. The written procedures should provide reasonable assurance that only appropriately trained personnel will handle and use licensed material without undue hazard to workers or members of the public.

General Safety Procedures

The written procedures should include the following elements:

- Contamination controls,

- Waste disposal practices,

- Personnel and area monitoring (including limits),

- Use of protective clothing and equipment,

- Safe handling of radioactive materials,

- Recording requirements,

- Reporting requirements, and

- Responsibilities.

These procedures should include policies for:

- Frequency of personnel monitoring,

- Performing molybdenum-99 breakthrough measurements of the first eluate after receipt of a molybdenum-99/technetium-99m generator,

- Use of appropriate shielding (See Figure 8.8),

- Frequent glove changes to minimize exposure to the individual and to avoid spread of contamination in the facility, and

- Special procedures for higher risk activities, such as use of radioiodine and repair of chemistry synthesis equipment for PET radiopharmaceuticals.

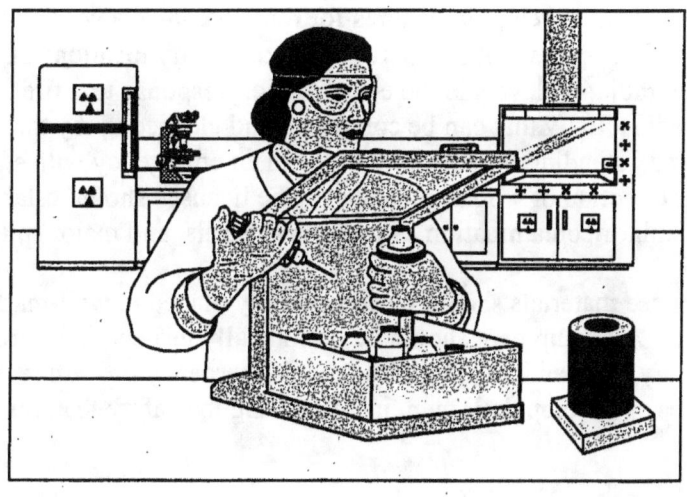

1556-059varb.ppt
091406

Figure 8.8 Use of Appropriate Shielding.

Applicants should also develop radionuclide-specific procedures based on the respective hazards associated with the radionuclides. General safety guidelines are described in Appendix Q. Applicants should use these guidelines to aid in the development of their own procedures for the safe use of radionuclides.

Furthermore, applicants that produce radioactive materials using an accelerator should also refer to the safety procedures found in NUREG-1556, Vol. 21, "Consolidated Guidance About Materials Licenses: Program-Specific Guidance About Possession Licenses for Production of Radioactive Materials Using an Accelerator."

Licensees should determine if they have areas that require posting in accordance with 10 CFR 20.1902, unless they meet the exemptions listed in 10 CFR 20.1903. Also, containers of licensed material (including radioactive waste) must be labeled in accordance with 10 CFR 20.1904, unless they meet the exemptions in 10 CFR 20.1905.

Emergency Procedures

Accidents and emergencies can happen during any operation with radionuclides, including their receipt, transportation, use, transfer, and disposal. Such incidents can result in contamination or release of material to the environment and unintended radiation exposure to workers and members of the public. In addition, loss or theft of licensed material, and fires involving radioactive material, can adversely affect the safety of personnel and members of the public. Applicants should therefore develop and implement procedures to minimize, to the extent practical, the potential impact of these incidents on personnel, members of the public, and the environment.

Applicants should establish written procedures to handle events ranging from a minor spill to a major accident that may require intervention by outside emergency response personnel. These procedures should include provisions for immediate response, after-hours notification, handling of each type of emergency, equipment, and the appropriate roles of staff and the RSO. In addition, the licensee should develop procedures for routine contacts with its local fire department officials to inform them of its operations and identify locations of radioactive materials and elevated radiation levels in the event of their response to a fire. Except for minor spills or releases of radioactivity that can be controlled and cleaned up by the user, licensee staff should have a clear understanding of their limitations in an emergency with step-by-step instructions and clear direction of whom to contact. The licensee should establish clear delineations between minor contamination events, minor spills, and major spills and events.

Emergency spill response materials should be strategically placed in well-marked locations for use by all trained staff. All equipment should be periodically inspected for proper operation and replenished as necessary. Appendix Q includes model emergency procedures. Applicants may adopt these procedures or develop their own, incorporating the safety features included in these model procedures.

Certain incidents and emergencies require notification of NRC. Appendix T provides a list of major NRC reporting and notification requirements relevant to commercial radiopharmacies.

Response from Applicant: Submit the following statements:

"We have developed and will implement and maintain written procedures for the safe use of radioactive materials that address:

- Facility and personnel radioactive contamination minimization, detection, and control;

- Performing molybdenum-99 breakthrough measurements on the first eluate after receipt of the molybdenum-99/technetium-99m generator; and

- Use of protective clothing and equipment by personnel

that meet the requirements in 10 CFR 20.1101, 10 CFR 20.1801, 10 CFR 20.1802, 10 CFR 30.34(g), and 10 CFR 19.11(a)(3), as applicable";

AND

"We have developed and will implement and maintain written procedures for identifying and responding to emergencies involving radioactive material, including:

- Lost, stolen, or missing licensed material;

- Exposures to personnel and the public in excess of NRC regulatory limits;

- Releases of licensed materials in effluents and the sanitary sewer in excess of NRC regulatory limits;

- Excessive radiation levels or radioactive material concentrations in restricted or unrestricted areas;

- Radioactive spills and contamination;

- Fires, explosions, and other disasters with the potential for the loss of containment of licensed material; and

- Routine contacts with local fire departments and local law enforcement agencies (LLEA),

that meet the requirements in 10 CFR 20.1101, 10 CFR 20.2201-2203, and 10 CFR 30.50, and other requirements, as applicable."

8.10.7 SURVEYS

Regulations: 10 CFR 30.53, 10 CFR 20.1501, 10 CFR 20.2103.

Criteria: Licensees are required to make surveys of potential radiological hazards in their workplace. Records of survey results must be maintained.

Discussion: Surveys are evaluations of radiological conditions and potential hazards (see Figure 8.9). These evaluations may be measurements (e.g., radiation levels measured with survey instrument or results of wipe tests for contamination), calculations, or a combination of measurements and calculations. The selection and proper use of appropriate instruments is one of the most important factors in ensuring that surveys accurately assess the radiological conditions. In order to meet regulatory requirements for surveying, measurements of radioactivity should be understood in terms of its properties (i.e., alpha, beta, gamma) and compared to the appropriate limits.

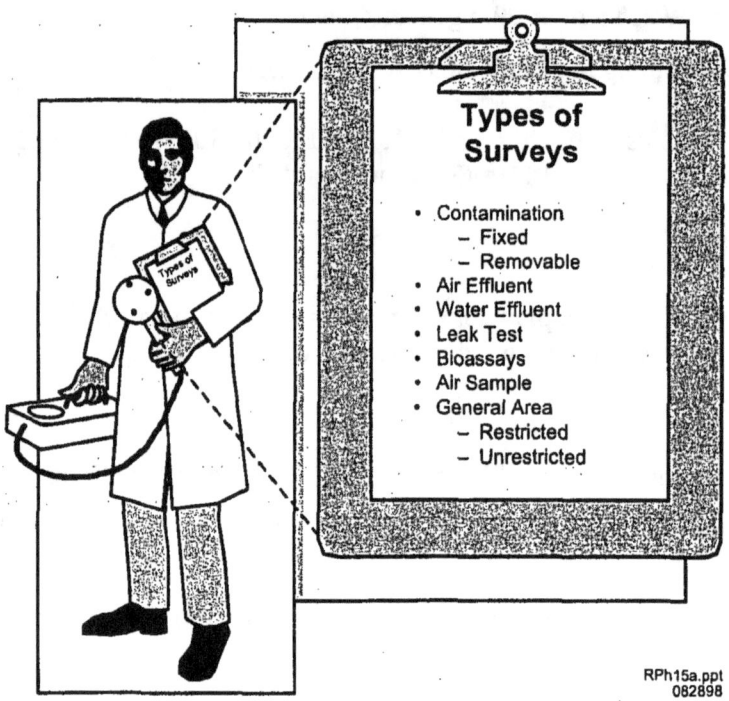

Figure 8.9 Types of Surveys. *There are many different types of surveys performed by radiopharmacy licensees.*

Radiation surveys are used to detect and evaluate contamination of:

- Facilities (restricted and unrestricted areas),

- Equipment,

- Incoming and outgoing radioactive packages, and

- Personnel (during use, transfer, or disposal of licensed material) (see Figure 8.10).

Surveying arm and hand using survey
meter and beta/gamma probe.

Surveying feet and legs using survey
meter and beta/gamma probe

RPh16a.ppt
082898

Figure 8.10 Personnel Surveys. *Users of unsealed licensed material should check themselves for contamination (frisk) before leaving the restricted areas within the radiopharmacy.*

Surveys are also used to plan work in areas where licensed material or radiation exists (see Figure 8.11) and to evaluate doses to workers and individual members of the public.

Surveys are required when it is reasonable under the circumstances to evaluate a radiological hazard and when necessary for the licensee to comply with the appropriate regulations. Many different types of surveys may need to be performed, due to the particular use of licensed materials. The most important are as follows:

- Surveys for radioactive contamination that could be present on surfaces of floors, walls, laboratory furniture, and equipment;

- Measurements of radioactive material concentrations in air for areas where radiopharmaceuticals are handled or processed in unsealed form and where operations could expose workers to the inhalation of radioactive material (e.g., radioiodine) or where licensed material is or could be released to unrestricted areas;

- Bioassays to determine the kinds, quantities or concentrations, and in some cases, the location, of radioactive material in the human body. Radioiodine uptake in a worker's thyroid gland is commonly measured by external counting using a specialized thyroid detection probe;

- Surveys of external radiation exposure levels in both restricted and unrestricted areas; and

- Surveys of radiopharmaceutical packages entering (e.g., from suppliers and returns from customers) and departing (e.g., prepared radiopharmaceuticals for shipment to customers).

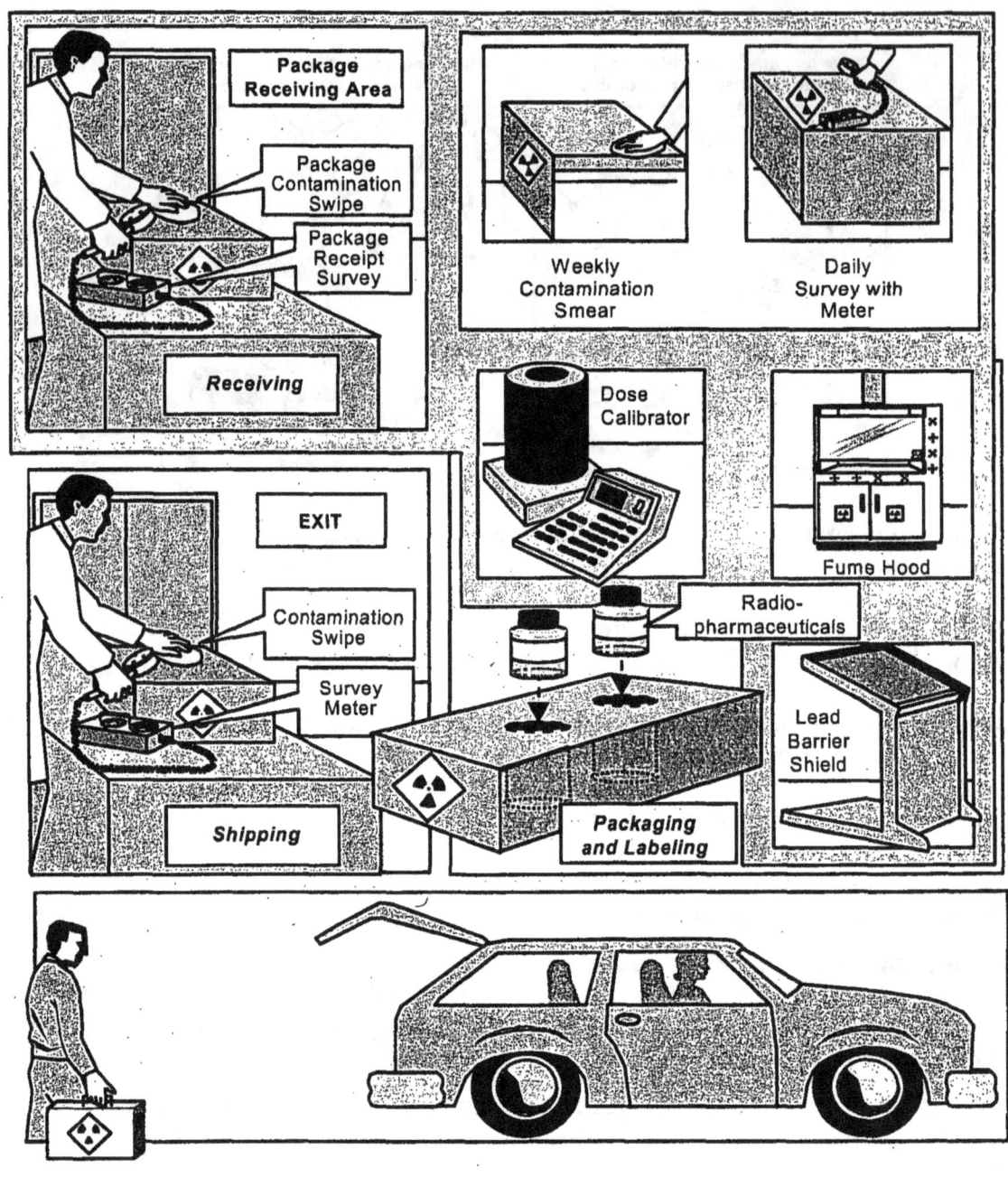

Figure 8.11   Typical Surveys at Radiopharmacy.

The frequency of routine surveys depends on the nature, quantity, and use of radioactive materials, as well as the specific protective facilities, equipment, and procedures that are designed to protect workers from external and internal exposure. Also, the frequency of the survey depends on the type of survey, such as those listed above. Appendix R, "Radiation Surveys," contains a model procedure for radiation survey frequencies.

Not all instruments can measure a given type of radiation (e.g., alpha, beta, and gamma). The presence of other radiation may interfere with a detector's ability to measure the radiation of interest. The energy of the radiation may not be high enough to penetrate some detector windows and be counted. The correct selection, calibration, and use of radiation detection instruments is an important aspect of any Radiation Safety Program.

Although 10 CFR Part 20 does not specify limits for surface contamination, it does specify dose limits for unrestricted areas (2 millirem in any 1 hour) and posting requirements (5 millirem in any one hour for "Radiation Areas"). Applicants should propose and justify their removable surface contamination and radiation level action limits that will require action to (1) reduce the contamination or radiation level; or (2) institute additional restrictions on access to the area. See Table R.1 in Appendix R for guidance on surface contamination limits acceptable to NRC.

Undetected Contamination and Loss of Control of Licensed Material

Due to the large quantities of licensed material in liquid form often handled by radiopharmacy personnel, there can be a greater potential for radioactive material contamination. Radiation surveys, if properly conducted as outlined in this section, will normally detect contamination before it leaves the licensee's restricted area (e.g., radiopharmaceutical preparation and packaging areas). If detected within the restricted area during or shortly following radiopharmaceutical preparation, the licensee can normally complete standard decontamination activities to mitigate the spread of the contamination outside the restricted area.

There have been several instances involving NRC licensees, including radiopharmacies, in which contamination has not been detected (usually due to no survey, or an inadequate survey, being performed) and has been inadvertently removed from the restricted area. Typically the contamination has been deposited on an outgoing package containing radioactive material, the skin or clothing of a licensee employee leaving the facility, or both. Once the contamination leaves the licensee's restricted area, control of the radioactive material is lost. At this point, the contamination has a high probability of reaching public locations outside the radiopharmacy, including one or more of its customers (e.g., a hospital). Contamination incidents such as this can create public health, regulatory, and public relations problems for licensees. In virtually all cases, the events could have been avoided if licensee personnel had performed an adequate radiation survey to detect the contamination before it left the restricted area. NRC Information Notice 98-18, "Recent Contamination Incidences Resulting From Failure to Perform Adequate Surveys," dated May 13, 1998, describes some such incidents involving NRC licensees, followed by a summary of NRC requirements to perform adequate and timely surveys.

Response from Applicant: Submit the following statement: "We have developed and will implement and maintain written procedures for a survey program that specifies the performance of radiation and contamination level surveys in restricted and unrestricted areas, personnel contamination monitoring, action levels, and the frequencies and records maintenance of those surveys and monitoring that meet the requirements in 10 CFR 30.53, 10 CFR 20.1501, and 10 CFR 20.2103, as applicable."

References: NRC Information Notice 98-18, "Recent Contamination Incidences Resulting From Failure to Perform Adequate Surveys," dated May 13, 1998, can be found at http://www.nrc.gov/reading-rm/doc-collections/gen-comm/info-notices/1998/in98018.html.

8.10.8 DOSAGE MEASUREMENT SYSTEMS

Regulation: 10 CFR 32.72(c).

Criteria: Commercial radiopharmacy licensees must possess and use instrumentation capable of accurately measuring the radioactivity in radioactive drugs.

Discussion: Due to the potential for radiopharmacy errors to adversely affect their customers (medical facilities) and their customers' patients, each dosage of a radioactive drug must be measured prior to transfer to provide high confidence that the correct amount of the radioactive drug is transferred in accordance with the customer's request.

The applicant must have procedures for the use of the instrumentation, including the measurement, by direct measurement or by a combination of measurement and calculation, of the amount of radioactivity in dosages of alpha-, beta-, gamma-, or photon-emitting radioactive drugs prior to their transfer for commercial distribution.

These procedures must ensure that the dose calibrator, or other dose measurement system, functions properly. This is accomplished by performing periodic checks and tests prior to first use, followed by checks at specified intervals, and following repairs that could affect system performance. Equipment used to measure dosages that emit gamma, alpha, or beta radiation must be calibrated for the applicable radionuclide being measured. For photon-emitters, activity measurement is a fairly straightforward determination; however, for beta-emitters, a correction factor is often necessary to accurately determine the activity. There are inherent technical difficulties to overcome in the determination and application of beta-correction factors. These difficulties include dependence on geometry, lack of an industry standard for materials used in the manufacture of both vials and syringes, and lack of a National Institute of Standards and Technology (NIST)- traceable standard for all radionuclides currently in use. If radiopharmacies intend to initially distribute (i.e., measure, prepare, and label) beta-emitting radionuclides, the applicant must provide the calculation to demonstrate its ability to accurately dispense such materials. If the applicant intends to use beta-correction factors supplied by the instrument manufacturer, or other entity, it should include a means for ensuring the accuracy of the supplied factor. If radiopharmacy applicants intend to only redistribute beta-emitting radionuclides that have been previously prepared and distributed by other persons licensed pursuant to 10 CFR 32.72, then the correction factor calculation is not required.

Licensees must assay patient dosages in the same type of vial and geometry as used to determine the correct dose calibrator settings. The use of different vials or syringes may result in measurement errors, for example, due to the variation of bremsstrahlung created by interaction between beta particles and the differing dosage containers. Licensees are reminded that beta emitters should be shielded using a low-atomic-numbered material to minimize the production of bremsstrahlung, followed by a high-atomic-numbered material thick enough to attenuate the bremsstrahlung intensity.

For each dose measurement system, specific periodic tests must be performed, as appropriate to the system, to ensure correct operation. Typically, all systems must be checked each day of use for constancy to ensure continued proper operation of the system. In addition, other appropriate tests may include accuracy (for the range of energies to be measured), linearity (for the range of activities to be measured), and geometry dependence (for the range of volumes and product containers).

The applicant should ensure that it possesses a sufficient number of such instruments to allow for periods when instruments are out of service for repair and calibration.

Appendix O contains a model procedure for dose calibrator testing.

Response from Applicant: The applicant shall describe the types of systems (measurement or combination of measurement and calculation) it intends to use for the measurement of alpha-, beta-, gamma-, and photon-emitting radioactive drugs;

AND

For each dose measurement system used to measure the amount of radioactivity in alpha-, beta-, gamma-, or photon-emitting radioactive drugs, state: "We have developed, and will implement and maintain a written procedure for the performance of dose measurement system checks and tests that meets the requirements in 10 CFR 32.72(c)";

AND

If applicable, the applicant must include a sample calculation for determining beta-correction factors for dose calibrators with ionization chambers;

> Radiopharmacies that intend to initially distribute (i.e., measure, prepare, and label) beta-emitting radionuclides must provide the calculation to demonstrate its ability to accurately dispense such materials; however, a correction factor calculation is not required if radiopharmacy applicants intend to only redistribute beta-emitting radionuclides that were previously prepared and distributed by others who are licensed pursuant to 10 CFR 32.72.

OR

If applicable, the applicant must include a means for ensuring the accuracy of beta-correction factors supplied by the instrument manufacturer, or other entity.

8.10.9 TRANSPORTATION

Regulations: 10 CFR 71.5, 10 CFR 71.12, 10 CFR 71.13, 10 CFR 71.14, 10 CFR 71.47, 10 CFR 71.87, 49 CFR 107, 49 CFR 171-180, 49 CFR 390-397, 10 CFR 20.1101, 10 CFR 30.41, 10 CFR 30.51.

Criteria: Applicants who will prepare for shipment, ship, or transport radioactive materials, including radioactive waste, must develop, implement, and maintain safety programs for the transport of those materials to ensure compliance with NRC and DOT regulations.

Discussion: In accordance with a Memorandum of Understanding (MOU) between DOT and NRC, NRC inspects and enforces DOT's regulations governing the transport of radioactive materials by NRC's licensees.

The types and quantities of radioactive materials shipped by commercial radiopharmacy licensees will nearly always meet the criteria for shipment in a "Type A" package, as defined by DOT. The requirements for these packages include the provisions for shipping papers, packaging design standards, package marking and labeling, and radiation and contamination level limits. For radiopharmacies who transport their own packages, the packages must be blocked and braced, and shipping papers must be used and located properly in the driver's compartment.

Packaging used by commercial radiopharmacies typically includes military ammunition boxes, "briefcases," and cardboard/fiberboard boxes. These packages will normally meet the criteria for "Type A" quantities, which must meet specified performance standards to demonstrate that they will maintain the integrity of containment and shielding under normal conditions of transport. Such packages will normally withstand minor accident situations and rough handling conditions. The testing criteria for Type A packages are listed in 49 CFR 173.465. Before offering a Type A package for shipment, the shipper is responsible for ensuring that the package has been tested to meet the criteria for the contents and the configuration to be shipped and for maintaining a certificate of testing. Shippers are not required to personally test the packages but must ensure that the testing was performed before use and maintain a record of the testing.

The DOT regulations also require that individuals who perform functions related to the packaging and shipment of radioactive material packages receive training specific to those functions. The training must include a general awareness of DOT requirements, function-specific training for the individuals' duties, safety training, and security-awareness training. The DOT also specifies the frequency of the training and a record retention requirement for training (see Section 8.8.2).

An outline of DOT and NRC requirements generally relevant to commercial radiopharmacy operations is included for applicant and licensee reference in Appendix M.

Response from Applicant: No response is required. The licensee's program for transportation of radioactive materials will be reviewed during inspection.

8.10.10 MINIMIZATION OF CONTAMINATION

Regulations: 10 CFR 20.1406.

Criteria: Applicants for new licenses must describe how facility design and procedures for operation will minimize, to the extent practicable, contamination of the facility and the environment, facilitate eventual decommissioning, and minimize, to the extent practicable, the generation of radioactive waste.

Discussion: All applicants for new licenses need to consider the importance of designing and operating their facilities to minimize the amount of radioactive contamination generated at the site during its operating lifetime and to minimize the generation of radioactive waste during decontamination. In the case of commercial radiopharmacy applicants, these issues usually do not need to be addressed as a separate item, as they are included in responses to other items of the application.

The bulk of unsealed radioactive material utilized by radiopharmacies have short half-lives (under 120 days). These radionuclides do not pose a source of long-term contamination. Additionally, nearly all radioactive waste generated by radiopharmacies is stored for decay rather than transferred to a radioactive waste disposal facility.

The licensee may possess and redistribute sealed sources that contain radionuclides with long half-lives. These sealed sources have been approved by NRC or an Agreement State and, if used according to the respective SSDR certificate, usually pose little risk of contamination. Leak tests performed at the frequency specified in the SSDR certificate should identify defective sources. Leaking sources must be immediately withdrawn from use and decontaminated, repaired, or disposed of according to NRC requirements. These steps minimize the spread of contamination and reduce radioactive waste associated with decontamination efforts.

Response from Applicant: The applicant does not need to provide a response to this item under the following condition. NRC will consider that the above criteria have been met if the applicant's responses meet the criteria in the following sections: Facilities and Equipment; Radiation Safety Program - Safe Use of Radionuclides and Emergency Procedures; Radiation Safety Program - Surveys; Radiation Safety Program - Leak Testing; and Waste Management, of NUREG-1556, Vol. 13, Rev. 1, "Consolidated Guidance About Materials Licenses, Program-Specific Guidance About Commercial Radiopharmacies."

8.10.11 RADIOACTIVE DRUG LABELING FOR DISTRIBUTION

Regulations: 10 CFR 20.1901, 10 CFR 20.1904, 10 CFR 20.1905, 10 CFR 30.34(g), and 10 CFR 32.72(a)(4).

Criteria: The labels affixed to radioactive drugs for distribution must have the required color, symbol, and wording.

Discussion: The licensee must label each "transport radiation shield" to show the radiation symbol as described in 10 CFR 20.1901. The label must also include the words "CAUTION, RADIOACTIVE MATERIAL," or "DANGER, RADIOACTIVE MATERIAL," the name of the radioactive drug or its abbreviation, and the quantity of radioactivity at a specified date and time. The phrase "transport radiation shield" refers to the primary shield for the radioactive drug, which may include the syringe, vial, or syringe or vial shield. The "transport radiation shield" should be constructed of material appropriate for the isotope to be transferred for commercial distribution. The "transport radiation shield" does not refer to the outer suitcase, packaging, or other carrying device, even though that barrier may provide some radiation shielding.

The licensee must label each syringe, vial, or other container (e.g., generator or ampoule) used to hold radioactive drugs to be transferred for commercial distribution to show the radiation symbol, as described in 10 CFR 20.1901. The label must include the words "CAUTION, RADIOACTIVE MATERIAL," or "DANGER, RADIOACTIVE MATERIAL," and an identifier that ensures the syringe, vial, or other container can be correlated with the information on the "transport radiation shield" label. The identifier must provide a correlation between the syringe, vial, or other container and the information on the label of its "transport radiation shield." Identifiers may include the prescription number, the name of the radioactive drug or its abbreviation, the name of the patient, or the clinical procedure.

Response from Applicant: The applicant must:

- Describe all labels, indicating the colors to be used, that will accompany the products and describe where each label is placed (e.g., on the "transport radiation shield" or on the container used to hold the radioactive drug); and

- Agree to affix the required labels to all "transport radiation shields" and to each container used to hold the radioactive drugs.

8.10.12 RADIOACTIVE DRUG SHIELDING FOR DISTRIBUTION

Regulations: 10 CFR 32.72(a)(3), 10 CFR 20.1201, 10 CFR 20.1207, 10 CFR 20.1208.

Criteria: The shielding provided for each radioactive drug to be distributed must be adequate for safe handling and storage by the pharmacy's customers to maintain occupational exposures ALARA.

Discussion: The applicant must provide appropriate "transport radiation shields" for the primary container of each radioactive drug that it intends to distribute. The shielding must be adequate for the types and quantities of radioactive materials that the applicant intends to distribute. Typically, "transport radiation shields" used by radiopharmacies have included two-piece, shielded syringe and vial containers (or "pigs"). Pharmacies have used lead and tungsten shields for gamma-emitting materials and plexiglass inserts for beta-emitters.

As general guidelines, "transport radiation shields" for technetium-99m products have ensured surface radiation levels of not more than 0.03 millisievert per hour (mSv/hr) (3 mrem/hr), due to the ease of shielding the low-energy gamma emitted. For iodine-131, surface dose rates on "transport radiation shields" have been approved up to 0.5 mSv/hr (50 mrem/hr) for diagnostic dosages and up to 1.5 mSv/hr (150 mrem/hr) for therapeutic dosages. The applicant should select appropriate shielding materials and dimensions to ensure not only that occupational doses are ALARA, but also that the "transport radiation shield" can be easily handled.

Response from Applicant: For each radioactive drug to be distributed (except for products intended for redistribution without manipulation and in the manufacturer's original shipping package):

* Indicate the radionuclide and the maximum activity for each type of container (e.g., vial, syringe);

* Describe the type and thickness of the "transport radiation shield" provided for each type of container; and

* Indicate the maximum radiation level to be expected at the surface of each "transport radiation shield" when the radioactive drug container is filled with the maximum activity.

Note: It is not acceptable to state that the applicant will comply with DOT regulations. The dose-rate limits that DOT imposes apply to the surface of the package, not the surface of the "transport radiation shield."

8.10.13 LEAK TESTS

Regulations: 10 CFR 30.53, 10 CFR 20.1501, 10 CFR 20.2103.

Criteria: NRC requires testing to determine whether there is any radioactive leakage from the sealed sources. Records of the test results must be maintained.

Discussion: When issued, a license will require performance of leak tests at intervals approved by NRC or an Agreement State and specified in the SSDR certificate. The measurement of the leak test sample is a quantitative analysis requiring that instrumentation used to analyze the sample be capable of detecting 185 Bq (0.005 microcuries) of radioactivity.

Commercial radiopharmacies may have their sealed sources leak tested by an individual licensed, by NRC or an Agreement State, to perform leak testing, or radiopharmacies may perform leak testing of their own sealed sources. Appendix L contains a model procedure for performance of leak testing and sample analysis. If the radiopharmacy has its leak testing performed by a licensed leak test provider, the radiopharmacy is expected to take the leak test samples according to the sealed source manufacturer's and the leak test provider's kit instructions and return it to the provider for evaluation and reporting results. Leak test samples should be collected at the most accessible area where contamination would accumulate if the sealed source were leaking.

Applicants must specifically request authorization to perform leak testing as a service to other licensees. Requests to provide leak testing as a service to other licensees will be reviewed and, if approved, they will be authorized by NRC staff via a license condition.

Response from Applicant: Submit the following statement: "We have developed and will implement and maintain written procedures for leak testing that meet the requirements in 10 CFR 30.53, 10 CFR 20.1501, and 10 CFR 20.2103."

Note: Leak testing is authorized via a license condition.

8.11 ITEM 11: WASTE MANAGEMENT

Regulations: 10 CFR 20.2001(a), 10 CFR 20.2003, 10 CFR 20.2006, 10 CFR 20.1904(b), 10 CFR 20.2108, 10 CFR 30.51.

Criteria: Radioactive waste must be disposed of in accordance with regulatory requirements and license conditions. Appropriate records of waste disposal must be maintained.

Discussion: Radioactive waste is normally generated when conducting licensed activities. Such waste may include used or unused radioactive material, or unusable items contaminated with radioactive material (e.g., absorbent paper, gloves). Licensees may not receive radioactive waste from other licensees for processing, storage or disposal, unless specifically authorized to do so by NRC. Commercial radiopharmacies may request authorization to receive certain radioactive waste returned from their customers. For guidance on receiving radioactive waste from customers, refer to the Section titled, "Radiation Safety Program - Waste Management, Returned Wastes from Customers."

All radioactive waste must be stored in appropriate containers until its disposal and the integrity of the waste containers must be assured. Radioactive waste containers must be appropriately labeled. All radioactive waste must be secured against unauthorized access or removal. The NRC requires commercial radiopharmacy licensees to manage radioactive waste generated at their facilities by one or more of the following methods:

- Decay-in-Storage (DIS),
- Transfer to an authorized recipient, and
- Release into sanitary sewerage.

Licensees may chose any one or more of these methods to dispose of their radioactive waste. It has been NRC's experience that most commercial radiopharmacies dispose of radioactive waste by decay-in-storage because the majority of licensed materials used by these facilities have short half-lives.

An applicant's programs for management and disposal of radioactive waste should include procedures for handling, safe and secure storage, characterization, minimization, and disposal of radioactive waste. Appropriate training should be provided to waste handlers. Regulations require that licensees maintain all appropriate records of disposal of radioactive waste.

Disposal By Decay-in-Storage (DIS)

NRC permits licensed materials with half-lives of less than or equal to 120 days to be disposed of by DIS. Waste should be held in storage until the radiation exposure rate cannot be distinguished from background radiation levels. Applicants should assure that adequate space and facilities are available for the storage of such waste. Procedures for management of waste by DIS should include methods of segregation, surveys prior to disposal, and maintenance of records of disposal.

Licensees can minimize the need for storage space if radioactive waste is segregated according to physical half-life. Segregation of waste is accomplished by depositing radionuclides of shorter physical half-lives in containers separate from those used to store radioactive waste with longer physical half-lives. Radioactive waste with shorter half-lives will take less time to decay and thus may be disposed of in shorter periods of time, freeing storage space.

Used syringes/needles and vials returned from pharmacy customers (medical facilities) are considered both biohazardous and radioactive waste since these items may be contaminated with customer's patients' blood or other body fluids. Following completion of DIS, such waste may be disposed of as biohazardous waste (medical waste) if radiation surveys (performed in a low-background area and without any interposed shielding) of the waste at the end of the holding period indicate that radiation levels are indistinguishable from background radiation levels.

Radioactive material labels on the used syringes/needles cannot be defaced without exposing employees to the risk of injury from the needles. Additionally, exposing employees to the risk of injury from needles would place licensees in violation of the Occupational Safety and Health Administration regulations in 29 CFR 1910.1030(d)(1), which require precautions to prevent contact with blood or other potentially infectious materials, including recommendations not to manipulate used syringes/needles by hand. Thus, radiopharmacy licensees do not have to deface or remove radiation labels from individual containers and packages (e.g., syringes, vials) inside waste barrels/containers intended for disposal as medical waste, provided the following conditions are met:

- The radioactive material labels on the outer waste barrels/containers will be defaced or removed prior to transfer to a waste disposal firm;

- Waste barrels/containers are sealed prior to delivery to the waste disposal firm;

- Waste barrels/containers will be delivered directly from the licensee's facility to a waste disposal firm for disposal;

- Medical waste is incinerated, and not sent to a medical waste landfill; and

- The waste disposal firm is notified that the barrels/containers must not be opened at any point, and for any reason, prior to incineration.

Other pharmacy radioactive waste that has not been returned from customers and has not otherwise come into contact with blood or body fluids should not have a biohazardous component. Following completion of DIS and provided it has been stored separately from radioactive biohazardous waste and contains no other hazardous components (e.g., needles, hazardous chemicals), such waste may require disposal as ordinary trash if radiation surveys (performed in a low-background area and without any interposed shielding) of the waste at the end of the holding period indicate that radiation levels are indistinguishable from background radiation levels. All radiation labels must be defaced or removed from containers and packages prior to final disposal as ordinary trash. If the decayed waste is compacted, all labels that are visible in the compacted mass must also be defaced or removed.

Records of DIS should include the date when the waste was put in storage for decay, date of disposal, results of final survey before disposal as ordinary trash, results of the background survey, identification of the instrument used to perform the survey, and the signature or initials of the individual performing the survey.

Transfer to an Authorized Recipient

Licensees may transfer radioactive waste to an authorized recipient for disposal. It has been NRC's experience that most commercial radiopharmacies only dispose of radioactive wastes with half-lives greater than 120 days to authorized recipients (e.g., low-level radioactive waste disposal facilities). Since radiopharmacy licensees typically possess small quantities of these materials, the volume of materials disposed of in this manner would also be minimal, if any. Currently, radiopharmacies use this system for waste disposal infrequently; therefore, detailed guidance is not provided in this document on the specific requirements related to the transfer of wastes to authorized recipients for disposal.

> Because of the difficulties and costs associated with disposal of sealed sources, applicants should preplan the disposal. Applicants may want to consider contractual arrangements with the source supplier as part of a purchase agreement.

Release Into Sanitary Sewerage

Licensees may dispose of radioactive waste by release into sanitary sewerage if each of the following conditions are met:

- Material is readily soluble (or is easily dispersible biological material) in water;

- Quantity of licensed material that the licensee releases into the sewer each month averaged over the monthly volume of water released into the sewer does not exceed the concentration specified in 10 CFR Part 20, Appendix B, Table 3;

- If more than one radionuclide is released, the sum of the ratios of the average monthly discharge of a radionuclide to the corresponding limit in 10 CFR Part 20, Appendix B, Table 3 cannot exceed unity; and

- Total quantity of licensed material released into the sanitary sewerage system in a year does not exceed the limits specified in 10 CFR 20.2003(a)(4).

Licensees are responsible for demonstrating that licensed materials discharged into the sewerage system are indeed readily dispersible in water. NRC IN 94-07, "Solubility Criteria for Liquid Effluent Releases to Sanitary Sewerage Under the Revised 10 CFR Part 20," dated January 1994, provides the criteria for evaluating solubility of liquid waste.

Applicants should develop and implement procedures to ensure that all releases of radioactive waste into the sanitary sewerage, if any, meet the criteria stated in 10 CFR 20.2003. Licensees are required to maintain accurate records of all releases of licensed material into sanitary sewerage.

Response from Applicant: Submit the following statement: "We have developed and will implement and maintain written procedures for waste management that meet the requirements in 10 CFR 20.1904(b), 10 CFR 20.2001(a), 10 CFR 20.2003, 10 CFR 20.2006, 10 CFR 20.2108, 10 CFR 30.51, as applicable."

Note: DIS is authorized via a license condition.

References: See the Notice of Availability on the inside front cover of this report to obtain copies of Policy and Guidance Directive PG 94-05, "Updated Guidance on Decay-In-Storage," dated October 1994; Information Notice 94-07, "Solubility Criteria for Liquid Effluent Releases to Sanitary Sewerage Under the Revised 10 CFR 20," dated January 1994; and Information Notice 84-94, "Reconcentration of Radionuclides Involving Discharges into Sanitary Sewerage Systems Permitted Under 10 CFR 20.203 (now 10 CFR 20.2003)," dated December 1984. Information Notices are available at http://www.nrc.gov/reading-rm/doc-collections/gen-comm/info-notices.

8.11.1 RETURNED WASTES FROM CUSTOMERS

Regulations: 10 CFR 20.2001(a), 10 CFR 30.33, 10 CFR 71.5.

Criteria: Commercial radiopharmacies may receive radioactive waste from customers. This radioactive waste is limited to items that originated at the radiopharmacy and that contained (or contain) radioactive material delivered for customer use (e.g., pharmacy-supplied syringes and vials and their contents).

Discussion: Commercial radiopharmacy licenses contain a license condition that permits radioactive waste, consisting of pharmacy-supplied items, to be received from their customers. The customer may return, and the radiopharmacy may accept for disposal, only items originating at the radiopharmacy that contained or contain radioactive material. This is limited to pharmacy-supplied syringes and vials and their contents. It is *not* acceptable for customers to return items originating at their facilities that are contaminated with radioactive material supplied by the pharmacy (e.g., gloves, absorbent material, IV tubing, patient contaminated items) (See Figure 8.12). If an applicant wishes a broader authorization for radioactive waste retrieval, the applicant must apply for a separate license as a radioactive waste broker under the general provisions of 10 CFR 20.2001(b) and 10 CFR 30.33.

Figure 8.12 Returned Waste. *Only items that originated at the radiopharmacy (pharmacy-supplied syringes and vials and their contents) may be returned to the radiopharmacy for disposal.*

Radiopharmacy customers, who act as the shipper for returned materials, should be supplied with detailed written instructions on how to properly prepare and package radioactive waste for return to the radiopharmacy. These instructions should clearly indicate that only items that contained or contain radioactive materials supplied by the radiopharmacy may be returned. In addition, these instructions should be adequate to ensure that customers comply with DOT and NRC regulations for the packaging and transport of licensed materials and for the radiation safety of drivers/couriers. Since customers may return unused syringes and vials, which may contain significant quantities of licensed material, the radiopharmacy should also include in their instructions methods for determining that the activities of radionuclides returned to the pharmacy are "limited quantities," or otherwise ensure that customers prepare and offer packages for transport that meet NRC and DOT requirements if the packages contain greater than limited quantities of radioactive material. The radiopharmacy should also have written instructions for pharmacy staff to address pick-up, receipt, and disposal of the returnable radioactive waste. Appendix S contains a model procedure for the return of pharmacy radioactive wastes from customers.

If the pharmacy chooses to take the responsibility to act as the shipper for returned materials, the pharmacy must ensure that its customer follows DOT and NRC regulations for the packaging and transport of licensed materials and for the radiation safety of drivers/couriers in the return process.

Response from Applicant: Submit the following statement: "We have developed and will implement and maintain written procedures for customer return of pharmacy-supplied syringes and vials and their contents, to specify that:

• Only pharmacy-supplied syringes and vials and their contents may be returned to the pharmacy,

- Instructions will be provided to radiopharmacy customers for the proper preparation and packaging of the radioactive waste for return to the radiopharmacy, and

- Instructions will be provided to pharmacy staff for the pick-up, receipt, and disposal of the returned radioactive waste that meet the requirements in 10 CFR 20.2001(a), 10 CFR 30.33, and 10 CFR 71.5, as applicable."

Note: Retrieval, receipt, and disposal of pharmacy-supplied syringes and vials from customers are authorized via a license condition.

8.12 ITEM 12: FEES

The next two items on NRC Form 313 are to be completed on the form itself.

On NRC Form 313, enter the appropriate fee category from 10 CFR 170.31 and the amount of the fee enclosed with the application.

Direct all questions about NRC's fees or completion of Item 12 of NRC Form 313 to the Office of the Chief Financial Officer at NRC headquarters in Rockville, Maryland, (301) 415-7554. Information about fees may also be obtained by calling NRC's toll free number, (800) 368-5642 extension 415-7554. The e-mail address is fees@nrc.gov.

8.13 ITEM 13: CERTIFICATION

Individuals acting in a private capacity are required to date and sign NRC Form 313. Otherwise, representatives of the corporation or legal entity filing the application should date and sign NRC Form 313. Representatives signing an application must be authorized to make binding commitments and to sign official documents on behalf of the applicant. As discussed previously in "Management Responsibility," signing the application acknowledges management's commitment and responsibilities for the Radiation Protection Program. NRC will return all unsigned applications for proper signature.

Notes:

- It is a criminal offense to make a willful false statement or representation on applications or correspondence (18 U.S.C. 1001).

- When the application references commitments, those items become part of the licensing conditions and regulatory requirements.

9 AMENDMENTS AND RENEWALS TO A LICENSE

It is the licensee's obligation to keep the license current. If any of the information provided in the original application is to be modified or changed, the licensee must submit an application for a license amendment before the change takes place; however, in accordance with 10 CFR 32.72(b)(5), commercial radiopharmacy licensees may allow individuals not named on their licenses to work as ANPs, provided that the individuals meet the minimum training and experience requirements of 10 CFR 32.72(b)(2) or (4), and the licensee notifies NRC in writing, with the documentation specified in 10 CFR 32.72(b)(5), as applicable, no later than 30 days after the licensee allows the individual to work as an ANP. Also, to continue the license after its expiration date, the licensee must submit an application for a license renewal at least 30 days before the expiration date (10 CFR 2.109, 10 CFR 30.36(a)).

Applicants for license amendment or renewal should do the following:

- Use the most recent guidance in preparing an amendment or renewal request;

- Submit in duplicate, either an NRC Form 313 or a letter requesting amendment or renewal;

- Provide the license number and docket number;

- For renewals, provide a complete and up-to-date application if many outdated documents are referenced or there have been significant changes in regulatory requirements, NRC's guidance, the licensee's organization, or its Radiation Protection Program. Alternatively, describe clearly the exact nature of the changes, additions, and deletions; and

- If a renewal is requested, provide the appropriate fee.

> **Using the suggested wording of responses and committing to using the model procedures in this report will expedite NRC's review.**

10 APPLICATIONS FOR EXEMPTIONS

Regulations: 10 CFR 19.31, 10 CFR 20.2301, 10 CFR 30.11.

Criteria: Licensees may request exemptions to regulations. The licensee must demonstrate that the exemption is authorized by law, will not endanger life or property or the common defense and security, and is otherwise in the public interest.

Discussion: Various sections of NRC's regulations address requests for exemptions (e.g., 10 CFR 19.31, 10 CFR 20.2301, 10 CFR 30.11(a)). These regulations state that NRC may grant an exemption, acting on its own initiative or on an application from an interested person.

Exemptions are not intended to revise regulations, are not intended for large classes of license, and are generally limited to unique situations. Exemption requests must be accompanied by descriptions of the following:

- Exemption and justification for it,

- Proposed compensatory safety measures intended to provide a level of health and safety equivalent to the regulation for which the exemption is being requested, and

- Alternative methods for complying with the regulation and why compliance with the existing regulation is not feasible.

> **Until NRC has granted an exemption in writing, NRC expects strict compliance with all applicable regulations.**

11 TERMINATION OF ACTIVITIES

Regulations: 10 CFR 30.34(b), 10 CFR 30.35(g), 10 CFR 30.36(d), 10 CFR 30.36(g), 10 CFR 30.36(h), 10 CFR 30.36(j), 10 CFR 30.51(f).

Criteria: The licensee must do the following:

- Notify NRC, in writing, within 60 days of any of the following:

  — the expiration of its license;

  — a decision to cease licensed activities permanently at the entire site (regardless of contamination levels);

  — a decision to cease licensed activities permanently in any separate building or outdoor area, if they contain residual radioactivity that makes them unsuitable for release according to NRC requirements;

  — no principal activities having been conducted at the entire site under the license for a period of 24 months;

  — no principal activities having been conducted for a period of 24 months in any separate building or outdoor area, if it contains residual radioactivity making it unsuitable for release according to NRC requirements;

- Submit a decommissioning plan, if required by 10 CFR 30.36(g);

- Conduct decommissioning, as required by 10 CFR 30.36(h) and 10 CFR 30.36(j);

- Submit, to the appropriate NRC Regional Office, completed NRC Form 314, "Certificate of Disposition of Materials" (or equivalent information) and a demonstration that the premises are suitable for release for unrestricted use (e.g., results of final survey); and

- Before a license is terminated, send the records important to decommissioning to the appropriate NRC Regional Office. If licensed activities are transferred or assigned in accordance with 10 CFR 30.34(b), transfer records important to decommissioning to the new licensee.

Discussion: As noted in several instances discussed in "Criteria," before a licensee can decide whether it must notify NRC, the licensee must determine whether residual radioactivity is present and, if so, whether the levels make the building or outdoor area unsuitable for release, according to NRC requirements. A licensee's determination that a facility is not contaminated is subject to verification by NRC inspection.

For guidance on the disposition of licensed material, see Section 8.11 on "Waste Management." For guidance on decommissioning records, see Section 8.5.2 on "Radioactive Materials - Financial Assurance and Recordkeeping for Decommissioning."

Response from Applicant: The applicant's obligations in this matter begin when the license expires or at the time the licensee ceases operations, whichever is earlier. These obligations are to undertake the necessary decommissioning activities, to submit NRC Form 314 or equivalent information, and to perform any other actions as summarized in the Criteria. The applicant is not required to submit a response to NRC during the initial application.

Reference: NRC Form 314, "Certificate of Disposition of Materials," is available at http://www.nrc.gov/reading-rm/doc-collections/forms.

APPENDIX A

United States Nuclear Regulatory Commission Form 313

NRC FORM 313 (10-2005) 10 CFR 30, 32, 33, 34, 35, 36, 39, and 40 **APPLICATION FOR MATERIAL LICENSE**	U.S. NUCLEAR REGULATORY COMMISSION	APPROVED BY OMB: NO. 3150-0120 EXPIRES: 10/31/2008 Estimated burden per response to comply with this mandatory collection request: 4.4 hours. Submittal of the application is necessary to determine that the applicant is qualified and that adequate procedures exist to protect the public health and safety. Send comments regarding burden estimate to the Records and FOIA/Privacy Services Branch (T-5 F53), U.S. Nuclear Regulatory Commission, Washington, DC 20555-0001, or by internet e-mail to infocollects@nrc.gov, and to the Desk Officer, Office of Information and Regulatory Affairs, NEOB-10202, (3150-0120), Office of Management and Budget, Washington, DC 20503. If a means used to impose an information collection does not display a currently valid OMB control number, the NRC may not conduct or sponsor, and a person is not required to respond to, the information collection.

INSTRUCTIONS: SEE THE APPROPRIATE LICENSE APPLICATION GUIDE FOR DETAILED INSTRUCTIONS FOR COMPLETING APPLICATION. SEND TWO COPIES OF THE ENTIRE COMPLETED APPLICATION TO THE NRC OFFICE SPECIFIED BELOW.

APPLICATION FOR DISTRIBUTION OF EXEMPT PRODUCTS FILE APPLICATIONS WITH: DIVISION OF INDUSTRIAL AND MEDICAL NUCLEAR SAFETY OFFICE OF NUCLEAR MATERIALS SAFETY AND SAFEGUARDS U.S. NUCLEAR REGULATORY COMMISSION WASHINGTON, DC 20555-0001 ALL OTHER PERSONS FILE APPLICATIONS AS FOLLOWS: IF YOU ARE LOCATED IN: ALABAMA, CONNECTICUT, DELAWARE, DISTRICT OF COLUMBIA, FLORIDA, GEORGIA, KENTUCKY, MAINE, MARYLAND, MASSACHUSETTS, MISSISSIPPI, NEW HAMPSHIRE, NEW JERSEY, NEW YORK, NORTH CAROLINA, PENNSYLVANIA, PUERTO RICO, RHODE ISLAND, SOUTH CAROLINA, TENNESSEE, VERMONT, VIRGINIA, VIRGIN ISLANDS, OR WEST VIRGINIA, SEND APPLICATIONS TO: LICENSING ASSISTANCE TEAM DIVISION OF NUCLEAR MATERIALS SAFETY U.S. NUCLEAR REGULATORY COMMISSION, REGION I 475 ALLENDALE ROAD KING OF PRUSSIA, PA 19406-1415	IF YOU ARE LOCATED IN: ILLINOIS, INDIANA, IOWA, MICHIGAN, MINNESOTA, MISSOURI, OHIO, OR WISCONSIN, SEND APPLICATIONS TO: MATERIALS LICENSING BRANCH U.S. NUCLEAR REGULATORY COMMISSION, REGION III 2443 WARRENVILLE ROAD, SUITE 210 LISLE, IL 60532-4352 ALASKA, ARIZONA, ARKANSAS, CALIFORNIA, COLORADO, HAWAII, IDAHO, KANSAS, LOUISIANA, MONTANA, NEBRASKA, NEVADA, NEW MEXICO, NORTH DAKOTA, OKLAHOMA, OREGON, PACIFIC TRUST TERRITORIES, SOUTH DAKOTA, TEXAS, UTAH, WASHINGTON, OR WYOMING, SEND APPLICATIONS TO: NUCLEAR MATERIALS LICENSING BRANCH U.S. NUCLEAR REGULATORY COMMISSION, REGION IV 611 RYAN PLAZA DRIVE, SUITE 400 ARLINGTON, TX 76011-4005

PERSONS LOCATED IN AGREEMENT STATES SEND APPLICATIONS TO THE U.S. NUCLEAR REGULATORY COMMISSION ONLY IF THEY WISH TO POSSESS AND USE LICENSED MATERIAL IN STATES SUBJECT TO U.S. NUCLEAR REGULATORY COMMISSION JURISDICTIONS.

1. THIS IS AN APPLICATION FOR (Check appropriate item) ☐ A. NEW LICENSE ☐ B. AMENDMENT TO LICENSE NUMBER _____ ☐ C. RENEWAL OF LICENSE NUMBER _____	2. NAME AND MAILING ADDRESS OF APPLICANT (Include ZIP code)
3. ADDRESS WHERE LICENSED MATERIAL WILL BE USED OR POSSESSED	4. NAME OF PERSON TO BE CONTACTED ABOUT THIS APPLICATION TELEPHONE NUMBER

SUBMIT ITEMS 5 THROUGH 11 ON 8-1/2 x 11" PAPER. THE TYPE AND SCOPE OF INFORMATION TO BE PROVIDED IS DESCRIBED IN THE LICENSE APPLICATION GUIDE.

5. RADIOACTIVE MATERIAL. a. Element and mass number; b. chemical and/or physical form; and c. maximum amount which will be possessed at any one time.	6. PURPOSE(S) FOR WHICH LICENSED MATERIAL WILL BE USED.
7. INDIVIDUAL(S) RESPONSIBLE FOR RADIATION SAFETY PROGRAM AND THEIR TRAINING EXPERIENCE.	8. TRAINING FOR INDIVIDUALS WORKING IN OR FREQUENTING RESTRICTED AREAS.
9. FACILITIES AND EQUIPMENT.	10. RADIATION SAFETY PROGRAM.
11. WASTE MANAGEMENT:	12. LICENSE FEES (See 10 CFR 170 and Section 170.31) FEE CATEGORY AMOUNT ENCLOSED $

13. CERTIFICATION. (Must be completed by applicant) THE APPLICANT UNDERSTANDS THAT ALL STATEMENTS AND REPRESENTATIONS MADE IN THIS APPLICATION ARE BINDING UPON THE APPLICANT.

THE APPLICANT AND ANY OFFICIAL EXECUTING THIS CERTIFICATION ON BEHALF OF THE APPLICANT, NAMED IN ITEM 2, CERTIFY THAT THIS APPLICATION IS PREPARED IN CONFORMITY WITH TITLE 10, CODE OF FEDERAL REGULATIONS, PARTS 30, 32, 33, 34, 35, 36, 39, AND 40, AND THAT ALL INFORMATION CONTAINED HEREIN IS TRUE AND CORRECT TO THE BEST OF THEIR KNOWLEDGE AND BELIEF.

WARNING: 18 U.S.C. SECTION 1001 ACT OF JUNE 25, 1948 62 STAT. 749 MAKES IT A CRIMINAL OFFENSE TO MAKE A WILLFULLY FALSE STATEMENT OR REPRESENTATION TO ANY DEPARTMENT OR AGENCY OF THE UNITED STATES AS TO ANY MATTER WITHIN ITS JURISDICTION.

CERTIFYING OFFICER – TYPED/PRINTED NAME AND TITLE	SIGNATURE	DATE

FOR NRC USE ONLY

TYPE OF FEE	FEE LOG	FEE CATEGORY	AMOUNT RECEIVED $	CHECK NUMBER	COMMENTS
APPROVED BY				DATE	

NRC FORM 313 (10-2005)   PRINTED ON RECYCLED PAPER

APPENDIX B

List of Documents Considered in Development of This NUREG

List of Documents Considered in Development of This NUREG

This report incorporates and updates the guidance previously found in the NUREG reports, Regulatory Guides (RGs), Policy and Guidance Directives (P&GDs), and Information Notices (INs) listed below. Other NRC documents, such as Manual Chapters (MCs), Inspection Procedures (IPs), and Memoranda of Understanding (MOU) were also consulted during the preparation of this report. The documents marked in Table B.1 with an asterisk (*) have been superseded and should not be used.

Table B.1   List of NUREG Reports, Regulatory Guides, and Policy and Guidance Directives

Document Identification	Title	Date
Draft RG DG-0006*	Guide for the Preparation of Applications for Commercial Nuclear Pharmacy Licenses	3/97
FC 410-4*	Guide for the Preparation of Applications for Nuclear Pharmacy Licenses	8/85
SRP 85-14*	Standard Review Plan for Applications for Nuclear Pharmacy Licenses	8/85
P&GD FC 86-9*	Authorizing Possession and Use of Depleted Uranium as Shielding for High Activity Molybdenum-99/Technetium-99m Generators	6/86
IN 89-25, Rev. 1	Unauthorized Transfer of Ownership or Control of Licensed Activities	12/94
IN 97-03	Defacing Labels to Comply with 10 CFR 20.1904(b)	2/97
IN 98-18	Recent Contamination Incidences Resulting from Failure to Perform Adequate Surveys	5/98
GL 95-09	Monitoring and Training of Shippers and Carriers of Radioactive Materials	11/95

APPENDIX C

Suggested Format for Providing Information Requested in Items 5 through 11 of NRC Form 313

Suggested Format for Providing Information Requested in Items 5 through 11 on NRC Form 313

Item No.	Title and Criteria	Yes	Description Attached
5.	**RADIOACTIVE MATERIAL** **Sealed And/Or Unsealed Byproduct Material** For unsealed materials: • Identify each radionuclide (element name and mass number) that will be used, the form, and the maximum requested possession limit.		☐
	AND		
	For potentially volatile materials (e.g., iodine-131): • Specify whether the material will be manipulated at the radiopharmacy.	☐	N/A
	For sealed sources and discrete sources of radium-226: • Identify each radionuclide (element name and mass number) that will be used in each source;		☐
	• Provide the manufacturer's (distributor's) name and model number for each sealed source and device and discrete source of radium-226 requested;		☐
	• We confirm that each sealed source, device, source/device combination, and discrete source of radium-226 is registered as an approved sealed source, device or discrete source by NRC or by an Agreement State;	☐	N/A
	• We confirm that the activity per source and/or device and its maximum activity will not exceed the maximum activity listed on the approved certificate of registration issued by NRC or by an Agreement State; and	☐	N/A
	• If the above information cannot be provided for the discrete source of radium-226, describe the discrete source and its physical boundaries.		☐
	For depleted uranium, specify the total amount (in kilograms).		

Item No.	Title and Criteria	Yes	Description Attached
5.	**RADIOACTIVE MATERIAL** *(Cont.)* **Financial Assurance and Recordkeeping for Decommissioning** If financial assurance is required, submit documentation required by 10 CFR 30.35.		☐
6.	**PURPOSE(S) FOR WHICH LICENSED MATERIAL WILL BE USED** For radiopharmaceuticals:		
	• We confirm that radiopharmaceuticals will be prepared under the supervision of an ANP or will be obtained from a supplier authorized pursuant to 10 CFR 32.72; and	☐	
	• Describe all licensed material to be distributed or redistributed.		☐
	For generators:		
	• We confirm that the generators will be obtained from a manufacturer licensed pursuant to 10 CFR 32.72, or under equivalent Agreement State requirements; and	☐	
	• We confirm that unused generators will be redistributed without opening or altering the manufacturer's packaging.	☐	
	For redistribution of used generators:		
	• Describe the procedures and instructions for safely repackaging the generators, including the use of the manufacturer's original packaging and minimization of migration of radioactive fluids out of the generator during transport;		☐
	• We confirm that the manufacturer's packaging and labeling will not be altered;	☐	
	• We confirm that the generator will not be distributed beyond the expiration date shown on the generator label;	☐	
	• We confirm that the redistributed generator will be accompanied by the manufacturer-supplied leaflet or brochure that provides radiation safety instructions for handling and using the generator; and	☐	
	• We confirm that only generators used in accordance with the manufacturer's instructions will be redistributed.	☐	

Item No.	Title and Criteria	Yes	Description Attached
6.	**PURPOSE(S) FOR WHICH LICENSED MATERIAL WILL BE USED** *(Cont.)*		
	For Redistribution of Sealed Sources — for Brachytherapy or Diagnosis:		
	• We confirm that the sealed sources for brachytherapy or diagnosis to be redistributed will be obtained from a manufacturer authorized to distribute sealed sources for brachytherapy or diagnosis in accordance with a specific license issued pursuant to 10 CFR 32.74 or under equivalent Agreement State requirements; and	☐	
	• We confirm that the manufacturer's packaging, labeling, and shielding will not be altered and that redistributed sources will be accompanied by the manufacturer-supplied package insert, leaflet, brochure, or other document that provides radiation safety instructions for handling and storing the sources.	☐	
	For Redistribution of Calibration and Reference Sealed Sources:		
	• We confirm that calibration and reference sealed sources to be redistributed to medical use licensees will be obtained from a person licensed pursuant to 10 CFR 32.74 to initially distribute such sources; and	☐	
	• We confirm that the manufacturer's labeling and packaging will not be altered and that redistributed sources will be accompanied by the manufacturer-supplied calibration certificate and the leaflet, brochure, or other document that provides radiation safety instructions for handling and storing the sources.	☐	
	For Redistribution of Prepackaged Units for *In Vitro* Tests:		
	• We confirm that the prepackaged units for *in vitro* tests to be redistributed will have been obtained from a manufacturer authorized to distribute the prepackaged units for *in vitro* tests in accordance with a specific license issued pursuant to 10 CFR 32.71 or under an equivalent license of an Agreement State.	☐	

Item No.	Title and Criteria	Yes	Description Attached
6.	**PURPOSE(S) FOR WHICH LICENSED MATERIAL WILL BE USED** *(Cont.)*		
	For Redistribution to General Licensees:		
	• We confirm that the manufacturer's packaging and labeling of the prepackaged units for *in vitro* tests will not be altered in any way; and	❏	
	• We confirm that each redistributed prepackaged unit for *in vitro* tests will be accompanied by the manufacturer-supplied package insert, leaflet, or brochure that provides radiation safety instructions for general licensees.	❏	
	For Redistribution to Specific Licensees:		
	• We confirm that the labels, package insert, leaflet, brochure, or other documents accompanying the redistributed prepackaged units for *in vitro* tests will NOT reference general licenses, exempt quantities, or NRC's regulations that authorize a general license (e.g., 10 CFR 31.11); and	❏	
	• We confirm that the labeling on redistributed prepackaged units for *in vitro* tests will conform to the requirements of 10 CFR 20.1901 and 20.1904.	❏	
	For Redistribution to Discrete Sources of radium-226:		
	• We confirm that the discrete sources of radium-226 will be obtained by a manufacturer authorized to distribute them.	❏	
	• We confirm that the manufacturer's packaging, labeling, and shielding will not be altered and that redistributed sources will be accompanied by the manufacture-supplied package insert, leaflet, brochure, or other document that provides radiation safety instructions for handling and storing sources.	❏	
	For radiopharmaceutical preparation, we will perform:		
	• compounding of iodine-131 capsules,	❏	
	• radioiodination,	❏	
	• chemical synthesis of PET radiopharmaceuticals,	❏	
	• technetium-99m kit preparation, and	❏	
	• other, specify.	❏	❏

Item No.	Title and Criteria	Yes	Description Attached
6.	**PURPOSE(S) FOR WHICH LICENSED MATERIAL WILL BE USED** *(Cont.)*		
	Supply specific information concerning the use of discrete sources of radium-226, sealed sources for reference and calibration, and depleted uranium.		☐
	We will provide customers the following radiation protection services involving licensed material:		
	• leak testing,	☐	☐
	• instrument calibration, and	☐	☐
	• other, specify.	☐	☐
7.	**INDIVIDUAL(S) RESPONSIBLE FOR RADIATION SAFETY PROGRAM AND THEIR TRAINING AND EXPERIENCE**		
	For applicant's management structure, provide:		
	• An organizational chart describing the management structure, reporting paths, and flow of authority between executive management and the RSO.		☐
	For the Radiation Safety Officer (RSO), provide:		
	• Name of the proposed RSO;		☐
	AND		
	A copy of the license (NRC or Agreement State) that authorized the uses requested and on which the individual was specifically named as the RSO, an ANP, or an AU;		☐
	OR		
	• Description of the training and experience demonstrating that the proposed RSO is qualified by training and experience applicable to commercial nuclear pharmacies.		☐
	Note: See Appendix G for convenient formats to use for documenting hours of training in basic radionuclide handling techniques and hours of experience using radionuclides.		
	For each proposed Authorized Nuclear Pharmacist (ANP), provide the following:		

Item No.	Title and Criteria	Yes	Description Attached
7.	**INDIVIDUAL(S) RESPONSIBLE FOR RADIATION SAFETY PROGRAM AND THEIR TRAINING AND EXPERIENCE** *(Cont.)*		
	• Name of the proposed ANP;		☐
	AND		
	• Pharmacist's license number and issuing entity;		☐
	AND		
	For an individual previously identified as an ANP on an NRC or Agreement State license or permit or by a commercial nuclear pharmacy that has been authorized to identify ANPs (10 CFR 32.72(b)(2)(i)):		
	• Previous license number (if issued by NRC) or a copy of the license (if issued by an Agreement State), or a copy of a permit issued by an NRC master materials licensee, a permit issued by an NRC or Agreement State broad-scope licensee, or a permit issued by an NRC Master Materials License broad-scope permittee on which the individual was named an ANP or a copy of an authorization as an ANP from a commercial nuclear pharmacy that has been authorized to identify ANPs.		☐
	OR		
	For an individual qualifying under 10 CFR 32.72(b)(4): • Documentation that the individual was a nuclear pharmacist preparing only radioactive drugs containing accelerator-produced radioactive material,		☐
	AND		
	• Documentation that the individual practiced at a pharmacy, a Government agency or Federally recognized Indian Tribe before November 30, 2007, or at all other pharmacies before August 8, 2009, or an earlier date as noticed by the NRC.		☐
	OR		

Item No.	Title and Criteria	Yes	Description Attached
7.	**INDIVIDUAL(S) RESPONSIBLE FOR RADIATION SAFETY PROGRAM AND THEIR TRAINING AND EXPERIENCE** *(Cont.)*		
	For an individual qualifying under 10 CFR 35.55(a):		
	• Copy of the certification(s) of the specialty board whose certification process has been recognized[2] under 10 CFR 35.55(a);		❐
	AND		
	• Written attestation, signed by a preceptor ANP, that training and experience required for certification have been satisfactorily completed and that a level of competency sufficient to function independently as an ANP has been achieved;		❐
	AND		
	• If applicable, a description of recent related continuing education and experience as required by 10 CFR 35.59;		❐
	OR		
	For an individual qualifying under 10 CFR 32.72(b)(2)(ii):		
	• Description of the training and experience specified in 10 CFR 35.55(b), demonstrating that the proposed ANP is qualified by training and experience;		❐
	AND		
	• Written attestation, signed by a preceptor ANP, that training and experience required for certification have been satisfactorily completed and that a level of competency sufficient to function independently as an ANP has been achieved;		❐
	AND		
	• If applicable, description of recent related continuing education and experience as required by 10 CFR 35.59.		❐

Item No.	Title and Criteria	Yes	Description Attached
7.	**INDIVIDUAL(S) RESPONSIBLE FOR RADIATION SAFETY PROGRAM AND THEIR TRAINING AND EXPERIENCE** *(Cont.)*		
	Notes:		
	• NRC Form 313A (ANP), "Authorized Nuclear Pharmacist Training and Experience and Preceptor Attestation [10 CFR 35.55]" may be used to document training and experience for those individuals qualifying under 10 CFR 35.55(a) or (b).		
	• Descriptions of training and experience will be reviewed using the criteria listed above. The NRC will review the documentation to determine if the applicable criteria in 10 CFR 32.72(b)(2), are met. If the training and experience do not appear to meet the criteria, the NRC may request additional information from the applicant or may request the assistance of the ACMUI in evaluating such training and experience.		
	For each proposed Authorized User (AU), provide the following:		
	• Name of each proposed AU;		☐
	AND		
	• Types, quantities, and proposed uses of licensed material;		☐
	AND		
	• A copy of the license (NRC or Agreement State) on which the individual was specifically named as an AU for the types, quantities, and proposed uses of licensed materials;		☐
	OR		
	• A copy of the permit maintained by a licensee of broad scope that identifies the individual as an AU for the types, quantities, and proposed uses of licensed materials;		☐
	OR		

Item No.	Title and Criteria	Yes	Description Attached
7.	**INDIVIDUAL(S) RESPONSIBLE FOR RADIATION SAFETY PROGRAM AND THEIR TRAINING AND EXPERIENCE** *(Cont.)* • Description of the training and experience demonstrating that the proposed AU is qualified by training and experience to use the requested licensed materials. The applicant may find it convenient to describe this training and experience using a format similar to Tables G-1 and G-2 in Appendix G.		☐
8.	**TRAINING FOR INDIVIDUALS WORKING IN OR FREQUENTING RESTRICTED AREAS (INSTRUCTIONS TO OCCUPATIONALLY EXPOSED WORKERS AND ANCILLARY PERSONNEL)** **Occupationally Exposed Workers and Ancillary Personnel** We have developed and will implement and maintain written procedures for a training program for each group of workers, including: topics covered; qualifications of the instructors; method of training; method for assessing the success of the training; and the frequency of training and refresher training.	☐	
	Personnel Involved in Hazardous Materials Package Preparation and Transport We have developed and will implement and maintain written procedures for training personnel involved in hazardous materials package preparation and transport that meet the requirements in 49 CFR 172.700, 49 CFR 172.702, and 49 CFR 172.704, as applicable.	☐	
	Instruction for Supervised Individuals Preparing Radiopharmaceuticals	**Need Not Be Submitted with Application**	
9.	**FACILITIES AND EQUIPMENT** Provide a copy of the registration or license from a State Board of Pharmacy as a pharmacy, or provide evidence that the facility is operating as a nuclear pharmacy within a Federal medical institution; **AND**	☐	☐

Item No.	Title and Criteria	Yes	Description Attached
9.	**FACILITIES AND EQUIPMENT** *(Cont.)*		
	Describe the facilities and equipment to be made available at each location where radioactive material will be used. For PET radiopharmacies, the description should include the method used to physically transfer licensed material to the different processes (e.g., chemical synthesis, dispensing). A diagram should be submitted that shows the applicant's entire facility and identifies activities conducted in all contiguous areas surrounding the facility. Diagrams should be drawn to a specified scale, or dimensions should be indicated.		❐
	Include the following information:		
	• Descriptions of the area(s) assigned for the production or receipt, storage, preparation, measurement, and distribution of radioactive materials and the location(s) for radioactive waste storage;		❐
	• Sufficient detail in the diagram to indicate locations of shielding, the proximity of radiation sources to unrestricted areas, and other items related to radiation safety;		❐
	• Descriptions of the ventilation systems, including gloveboxes or fume hoods, with pertinent airflow rates, area differential pressures, filtration equipment, and monitoring systems for the use or storage of radioactive materials likely to become airborne, such as compounding radioiodine capsules and dispensing radioiodine solutions; and,		❐
	• Verification that ventilation systems ensure that effluents are ALARA, are within the dose limits of 10 CFR 20.1301, and are within constraints for air emissions established under 10 CFR 20.1101(d)		❐
10.	**RADIATION SAFETY PROGRAM**		
	Audit Program		**Need Not be Submitted with Application**
	The applicant's program for reviewing the content and implementation of its Radiation Protection Program will be examined during inspections, but it should not be submitted in the license application.		

Item No.	Title and Criteria	Yes	Description Attached
10.	**RADIATION SAFETY PROGRAM** *(Cont.)*		
	Instruments		
	We will use equipment that meets the radiation monitoring instrument specifications and implement the model survey meter calibration program published in Appendix J to NUREG-1556, Vol. 13, Rev. 1, "Program-Specific Guidance About Commercial Radiopharmacy Licenses";	☐	
	OR		
	We will use equipment that meets the radiation monitoring instrument specifications published in Appendix J to NUREG-1556, Vol. 13, Rev. 1, "Program-Specific Guidance About Commercial Radiopharmacy Licenses," and instruments will be calibrated by other persons authorized by NRC, an Agreement State, or a licensing State to perform that service;	☐	
	OR		
	A description of alternative minimum equipment to be used for radiation monitoring and/or alternative procedures for the calibration of radiation monitoring equipment.		☐
	Material Receipt and Accountability		
	We have developed and will implement and maintain written procedures for safely opening packages that meet the requirements in 10 CFR 20.1906,	☐	
	AND		
	We will conduct physical inventories of sealed sources of licensed material at intervals not to exceed 6 months,	☐	
	AND		
	We have developed and will implement and maintain written procedures for licensed material accountability and control to ensure that: • license possession limits are not exceeded, • licensed material in storage is secured from unauthorized access or removal,	☐	

Item No.	Title and Criteria	Yes	Description Attached
10.	**RADIATION SAFETY PROGRAM** *(Cont.)*		
	• licensed material not in storage is maintained under constant surveillance and control, and		
	• records of receipt (either from the licensee's own production operations or from another licensee), transfer, and disposal of licensed material are maintained.		
	Occupational Dosimetry		
	We have developed and will implement and maintain written procedures for monitoring occupational dose that meet the requirements in 10 CFR 20.1501, 10 CFR 20.1502, 10 CFR 20.1201, 10 CFR 20.1202, 10 CFR 20.1203, 10 CFR 20.1204, 10 CFR 20.1207, 10 CFR 20.1208, and 10 CFR 20.2106, as applicable.	❏	
	Public Dose		**Need Not Be Submitted with Application**
	The applicant's program to control doses received by individual members of the public will be examined during inspection, but it should not be submitted in a license application.		
	Safe Use of Radionuclides and Emergency Procedures		
	We have developed and will implement and maintain written procedures for the safe use of radioactive materials that address:	❏	
	• facility and personnel radioactive contamination minimization, detection, and control;		
	• performing molybdenum-99 breakthrough measurements on the first eluate after receipt of the molybdenum-99/technetium-99m generator; and		
	• use of protective clothing and equipment by personnel		
	that meet the requirements in 10 CFR 20.1101, 10 CFR 20.1801, 10 CFR 20.1802, 10 CFR 30.34(g), and 10 CFR 19.11(a)(3), as applicable;		
	AND		
	We have developed and will implement and maintain written procedures for identifying and responding to emergencies involving radioactive material, including:	❏	
	• lost, stolen, or missing licensed material;		

Item No.	Title and Criteria	Yes	Description Attached
10.	**RADIATION SAFETY PROGRAM** *(Cont.)*		
	• exposures to personnel and the public in excess of NRC regulatory limits;		
	• releases of licensed materials in effluents and the sanitary sewer in excess of NRC regulatory limits;		
	• excessive radiation levels or radioactive material concentrations in restricted or unrestricted areas;		
	• radioactive spills and contamination;		
	• fires, explosions, and other disasters with the potential for the loss of containment of licensed material; and		
	• routine contacts with local fire departments and local law enforcement agencies		
	that meet the requirements in 10 CFR 20.1101, 10 CFR 20.2201-2203, and 10 CFR 30.50 and other requirements, as applicable.		
	Surveys		
	We have developed and will implement and maintain written procedures for a survey program that specifies the performance of radiation and contamination level surveys in restricted and unrestricted areas, personnel contamination monitoring, action levels, and the frequencies and records maintenance of those surveys and monitoring that meet the requirements in 10 CFR 30.53, 10 CFR 20.1501, and 10 CFR 20.2103, as applicable.	❐	
	Dosage Measurement Systems		
	Describe the types of systems (measurement or combination of measurement and calculation) to be used for the measurement of alpha-, beta-, gamma-, and photon-emitting radioactive drugs;		❐
	AND		
	For each dose measurement system used to measure the amount of radioactivity in alpha-, beta-, gamma-, or photon-emitting radioactive drugs, state: "We have developed and will implement and maintain a written procedure for the performance of dosage measurement system checks and tests that meets the requirement in 10 CFR 32.72©";	❐	

Item No.	Title and Criteria	Yes	Description Attached
10.	**RADIATION SAFETY PROGRAM** *(Cont.)*		
	AND		
	If applicable, include a sample calculation for determining beta-correction factors for dose calibrators with ionization chambers;		☐
	OR		
	If applicable, include a means for ensuring the accuracy of beta-correction factors supplied by the instrument manufacturer or other entity.		☐
	Transportation		**Need Not Be Submitted with Application**
	The applicant's program for transportation will be examined during inspection but should not be submitted in a license application.		
	Minimization of Contamination		
	The applicant does not need to provide a response to this item under the following condition: NRC will consider that the criteria have been met if the applicant's responses meet the criteria for the following sections of NUREG-1556, Vol. 13, Rev. 1: Section 8.9, "Item 9: Facilities and Equipment"; Section 8.10.6, "Safe Use of Radionuclides and Emergency Procedures"; Section 8.10.7, "Surveys"; Section 8.10.13, "Leak Tests"; and Section 8.11, "Item 11: Waste Management."		
	Radioactive Drug Labeling for Distribution		
	Describe all labels, indicating the colors to be used, that will accompany the products and describe where each label is placed (e.g., on the "transport radiation shield" or on the container used to hold the radioactive drug); and agree to affix the required labels to all "transport radiation shields" and to each container used to hold the radioactive drugs.		☐
	Radioactive Drug Shielding for Distribution		
	For each radioactive drug to be distributed (except for products intended for redistribution without manipulation and in the manufacturer's original shipping package), provide:		☐
	• The radionuclide and the maximum activity for each type of container (e.g., vial, syringe),		

Item No.	Title and Criteria	Yes	Description Attached
10.	**RADIATION SAFETY PROGRAM** *(Cont.)* • A description of the type and thickness of the "transport radiation shield" provided for each type of container, and • The maximum radiation level to be expected at the surface of each "transport radiation shield" when the radioactive drug container is filled with the maximum activity. **Leak Tests** We have developed and will implement and maintain written procedures for leak testing that meet the requirements in 10 CFR 30.53, 10 CFR 20.1501, and 10 CFR 20.2103.	❏	
11.	**WASTE MANAGEMENT** **Pharmacy-generated Radioactive Wastes** We have developed and will implement and maintain written procedures for waste management that meet the requirements in 10 CFR 20.2001(a), 10 CFR 20.2003, 10 CFR 20.2006, 10 CFR 20.2108, and 10 CFR 30.51, as applicable. **Returned Wastes from Customers** We have developed and will implement and maintain written procedures for customer return of pharmacy-supplied syringes and vials and their contents, to specify that: • only pharmacy-supplied syringes and vials and their contents may be returned to the pharmacy; • instructions will be provided to radiopharmacy customers for the proper preparation and packaging of the radioactive waste for return to the radiopharmacy; and • instructions will be provided to pharmacy staff for the pick-up, receipt, and disposal of the returned radioactive waste that meet the requirements in 10 CFR 20.2001(a), 10 CFR 30.33, and 10 CFR 71.5, as applicable.	❏ ❏	

APPENDIX D

Checklist for License Application

Checklist for License Application

D.1   ITEM 1: ACTION TYPE

ACTION TYPE:	ADMINISTRATIVE REVIEW:
☐ New ☐ Amendment ☐ Renewal	☐ Current Guidance Used ☐ References in Application Based on Current Regulations ☐ All Attachments Referenced Included ☐ Signature on Application

D.2   ITEM 2: LEGAL IDENTITY

NAME:	

D.3   ITEMS 2 & 3: ADDRESS

STORAGE & LOCATION-OF-USE ADDRESS:	MAILING ADDRESS:

D.4   ITEM 4: PERSON TO BE CONTACTED ABOUT THIS APPLICATION

CONTACT PERSON:	
TELEPHONE NUMBER:	

D.5 ITEMS 5 & 6: MATERIALS TO BE POSSESSED AND PROPOSED USES

Yes	No	Radionuclide	Form or Mfg/Model No.	Quantity	Purpose of Use	Specify Other Uses Not Listed on SSDR Certificate
		Byproduct Materials with Atomic No. 1-83	Any	_____ millicuries per nuclide, 1 curie total possession, except as noted:	10 CFR 32.72 and 10 CFR 30.41	❏ Not applicable ------------ ❏ Uses are:
		Molybdenum-99	Any	_____ curies	10 CFR 32.72 and 10 CFR 30.41	❏ Not applicable ------------ ❏ Uses are:
		Technetium-99m	Any	_____ curies	10 CFR 32.72 and 10 CFR 30.41	❏ Not applicable ------------ ❏ Uses are:
		Iodine-131	Any	_____ millicuries	10 CFR 32.72 and 10 CFR 30.41	❏ Not applicable ------------ ❏ Uses are:
		Fluorine-18	Any	_____ millicuries	10 CFR 32.72 and 10 CFR 30.41	❏ Not applicable ------------ ❏ Uses are:
		Iodine-123	Any	_____ millicuries	10 CFR 32.72 and 10 CFR 30.41	❏ Not applicable ------------ ❏ Uses are:
		Xenon-133	Any	_____ curies	10 CFR 32.72 and 10 CFR 30.41	❏ Not applicable ------------ ❏ Uses are:
		Any Byproduct Material in a Brachytherapy Source, as listed in 10 CFR 35.400	Sealed Sources	_____ millicuries	10 CFR 32.74 and 10 CFR 30.41	❏ Not applicable ------------ [ ] Uses are:

Yes	No	Radionuclide	Form or Mfg/Model No.	Quantity	Purpose of Use	Specify Other Uses Not Listed on SSDR Certificate
		Any Byproduct Material in a sealed source for diagnosis, as listed in 10 CFR 35.500	Sealed Sources	____curies per source and curies total	10 CFR 32.74 and 10 CFR 30.41	☐ Not applicable ------------ ☐ Uses are:
		Any byproduct material listed in 10 CFR 31.11(a)	Prepackaged units for *in vitro* diagnostic tests	____millicuries	10 CFR 31.11	☐ Not applicable ------------ ☐ Uses are:
		Any byproduct material authorized under 10 CFR35.65	Sealed Sources	____millicuries	Calibration and checking of the licensee's instruments and 10 CFR 32.74 and 10 CFR 30.41	☐ Not applicable ------------ ☐ Uses are:
		Depleted Uranium	Metal	____ kilograms	Shielding for molybdenum-99/technetium-99m generators	☐ Not applicable ------------ ☐ Uses are:
		Cesium-137	Sealed sources in compatible device as specified in Sealed Source and Device Registry Sheet	Not to exceed maximum activity per source as specified in Sealed Source and Device Registry Sheet	Instrument calibration	☐ Not applicable ------------ ☐ Uses are:
		Other (specify)				

D.6 ITEMS 7 THROUGH 11: TRAINING AND EXPERIENCE, FACILITIES AND EQUIPMENT, RADIATION SAFETY PROGRAM, AND WASTE DISPOSAL

Item Number and Title	Suggested Response	Yes	Description Attached
7. **Individual(s) Responsible for Radiation Safety Program and Their Training and Experience**	An organizational chart describing the management structure, reporting paths, and flow of authority between executive management and the RSO.		☐
7. **Individual(s) Responsible For Radiation Safety Program And Their Training And Experience** 7.1 **Radiation Safety Officer (RSO)** Name: _____	A copy of the license (NRC or Agreement State) that authorized the uses requested and on which the individual was specifically named as the RSO, an ANP, or an AU; <center>**OR**</center>Description of the training and experience demonstrating that the proposed RSO is qualified by training and experience as applicable to commercial nuclear pharmacies.	☐ ☐	☐ ☐
7. **Individual(s) Responsible for Radiation Safety Program and Their Training and Experience** 7.2 **Authorized Nuclear Pharmacist(s)** Name(s): _____	• Name of the proposed ANP; <center>**AND**</center>• Pharmacist's license number and issuing entity; <center>**AND**</center>*For an individual previously identified as an ANP on an NRC or Agreement State license or permit or by a commercial nuclear pharmacy that has been authorized to identify ANPs (10 CFR 32.72(b)(2)(i)):* • Previous license number (if issued by NRC) or a copy of the license (if issued by an Agreement State) or a copy of a permit issued by an NRC master materials licensee, a permit issued by an NRC or Agreement State broad-scope licensee, or a permit issued by an NRC Master Materials License broad-scope permittee on which the individual was named an ANP or a copy of an authorization as an ANP from a commercial nuclear pharmacy that has been authorized to identify ANPs. <center>**OR**</center>	☐ ☐ ☐	☐ ☐ ☐

Item Number and Title	Suggested Response	Yes	Description Attached
	For an individual qualifying under 10 CFR 32.72(b)(4):		
	• Documentation that the individual was a nuclear pharmacist preparing only radioactive drugs containing accelerator-produced radioactive material;	☐	☐
	AND		
	• Documentation that the individual practiced' at a pharmacy, a Government agency, or Federally recognized Indian Tribe before November 30, 2007, or at all other pharmacies before August 8, 2009, or an earlier date as noticed by the NRC.	☐	☐
	OR		
	For an individual qualifying under 10 CFR 35.55(a):		
	• Copy of the certification(s) of the specialty board whose certification process has been recognized under 10 CFR 35.55(a);	☐	☐
	AND		
	• Written attestation, signed by a preceptor ANP, that training and experience required for certification have been satisfactorily completed and that a level of competency sufficient to function independently as an ANP has been achieved;	☐	☐
	AND		
	• If applicable, description of recent related continuing education and experience as required by 10 CFR 35.59.	☐	☐
	OR		
	For an individual qualifying under 10 CFR 32.72(b)(2)(ii):		
	• Description of the training and experience specified in 10 CFR 35.55(b) demonstrating that the proposed ANP is qualified by training and experience;	☐	☐
	AND		
	• Written attestation, signed by a preceptor ANP, that training and experience required for certification have been satisfactorily completed and that a level of competency sufficient to function independently as an ANP has been achieved;	☐	☐

Item Number and Title	Suggested Response	Yes	Description Attached
	AND		
	• If applicable, description of recent related continuing education and experience as required by 10 CFR 35.59.	❏	❏
	Notes: • NRC Form 313A (ANP), "Authorized Nuclear Pharmacist Training and Experience and Preceptor Attestation [10 CFR 35.55]" may be used to document training and experience for those individuals qualifying under 10 CFR 35.55(a) or (b).		
	• Descriptions of training and experience will be reviewed using the criteria listed above. The NRC will review the documentation to determine if the applicable criteria in 10 CFR 32.72(b)(2) are met. If the training and experience do not appear to meet the criteria, the NRC may request additional information from the applicant or may request the assistance of the ACMUI in evaluating such training and experience.		
7. **Individual(s) Responsible for Radiation Safety Program and Their Training and Experience** 7.3 **Authorized User(s)** Name(s): _____	For each proposed AU: • Name of each proposed AU;	❏	❏
	AND		
	• Types, quantities, and proposed uses of licensed material;	❏	❏
	AND		
	• A copy of license (NRC or Agreement State) on which the individual was specifically named as an AU for the types, quantities, and proposed uses of licensed materials;	❏	❏
	OR		
	• A copy of the permit maintained by a licensee of broad scope that identifies the individual as an AU for the types, quantities, and proposed uses of licensed materials;	❏	❏
	OR		
	• Description of the training and experience demonstrating that the proposed AU is qualified by training and experience to use the requested licensed materials.	❏	❏

Item Number and Title	Suggested Response	Yes	Description Attached
8. **Training for Individuals Working in or Frequenting Restricted Areas** 8.1 **Occupationally Exposed Workers and Ancillary Personnel**	We have developed and will implement and maintain written procedures for a training program for each group of workers, including: topics covered, qualifications of the instructors, method of training, method for assessing the success of the training, and the frequency of training and refresher training.	☐	☐
8. **Training for Individuals Working in or Frequenting Restricted Areas** 8.2 **Training for Personnel Involved in Hazardous Materials Package Preparation and Transport**	We have developed and will implement and maintain written procedures for training personnel involved in hazardous materials package preparation and transport that meet the requirements in 49 CFR 172.700, 49 CFR 172.702, and 49 CFR 172.704, as applicable.	☐	☐
8. **Training for Individuals Working in or Frequenting Restricted Areas** 8.3 **Training for Supervised Individuals Preparing Radiopharmaceuticals**	The applicant's program for training of supervised individuals preparing radiopharmaceuticals will be examined during inspections, but should not be submitted in the license application.	N/A	
9. **Facilities and Equipment**	Provide a copy of the registration or license from a State Board of Pharmacy as a pharmacy, or provide evidence that the facility is operating as a nuclear pharmacy within a Federal medical institution; **AND**	☐	☐
	Describe the facilities and equipment to be made available at each location where radioactive material will be used, which includes the method used to physically transfer licensed material for PET radiopharmacies to the different processes (e.g., chemical synthesis, dispensing). A diagram should be submitted that shows the applicant's entire facility and identifies activities conducted in all contiguous areas surrounding the facility. Diagrams should be drawn to a specified scale, or dimensions should be indicated. Include the following information:	☐	☐
	• Descriptions of the area(s) assigned for the production or receipt, storage, preparation, measurement, and distribution of radioactive materials and the location(s) for radioactive waste storage;	☐	☐
	• Sufficient detail in the diagram to indicate locations of shielding, the proximity of radiation sources to unrestricted areas, and other items related to radiation safety;	☐	☐

Item Number and Title	Suggested Response	Yes	Description Attached
	• Descriptions of the ventilation systems, including gloveboxes or fume hoods, with pertinent airflow rates, area differential pressures, filtration equipment, and monitoring systems for the use or storage of radioactive materials that are likely to become airborne, such as compounding radioiodine capsules and dispensing radioiodine solutions; and,	☐	☐
	• Verification that ventilation systems ensure that effluents are ALARA, are within the dose limits of 10 CFR 20.1301, and are within the constraints for air emissions established under 10 CFR 20.1101(d).	☐	☐
10. Radiation Safety Program 10.1 Audit Program	The applicant's program for reviewing the content and implementation of its Radiation Protection Program will be examined during inspections, but it should not be submitted in the license application.		N/A
10. Radiation Safety Program 10.2 Radiation Monitoring Instruments	We will use equipment that meets the radiation monitoring instrument specifications and implement the model survey meter calibration program published in Appendix J to NUREG-1556, Vol. 13, Rev. 1, "Program-Specific Guidance About Commercial Radiopharmacy Licenses";	☐	
	OR		
	We will use equipment that meets the radiation monitoring instrument specifications published in Appendix J to NUREG-1556, Vol. 13, Rev. 1, "Program-Specific Guidance About Commercial Radiopharmacy Licenses," and instruments will be calibrated by persons authorized by NRC, an Agreement State, or a licensing State to perform that service;	☐	
	OR		
	Provide a description of alternative minimum equipment to be used for radiation monitoring and/or alternative procedures for the calibration of radiation monitoring equipment.	☐	☐
10. Radiation Safety Program 10.3 Material Receipt and Accountability	We have developed and will implement and maintain written procedures for safely opening packages that meet the requirements in 10 CFR 20.1906;	☐	
	AND		
	We will conduct physical inventories of sealed sources of licensed material at intervals not to exceed six months,	☐	
	AND		

Item Number and Title	Suggested Response	Yes	Description Attached
	We have developed and will implement and maintain written procedures for licensed material accountability and control to ensure that: • License possession limits are not exceeded, • Licensed material in storage is secured from unauthorized access or removal, • Licensed material not in storage is maintained under constant surveillance and control, and • Records of receipt, transfer, and disposal of licensed material are maintained.	❏	
10. Radiation Safety Program 10.4 Occupational Dosimetry	We have developed and will implement and maintain written procedures for monitoring occupational dose that meet the requirements in 10 CFR 20.1501, 10 CFR 20.1502, 10 CFR 20.1201, 10 CFR 20.1202, 10 CFR 20.1203, 10 CFR 20.1204, 10 CFR 20.1207, 10 CFR 20.1208, and 10 CFR 20.2106, as applicable.	❏	
10. Radiation Safety Program 10.5 Public Dose	The applicant's program to control doses received by individual members of the public will be examined during inspection, but it should not be submitted in a license application.		N/A
10. Radiation Safety Program 10.6 Safe Use of Radionuclides and Emergency Procedures	We have developed and will implement and maintain written procedures for the safe use of radioactive materials that address: • Facility and personnel radioactive contamination minimization, detection, and control; • Performing molybdenum-99 breakthrough measurements on the first eluate after receipt of the molybdenum-99/technetium-99m generator; and • Use of protective clothing and equipment by personnel that meet the requirements in 10 CFR 20.1101, 10 CFR 20.1801, 10 CFR 20.1802, 10 CFR 30.34(g), and 10 CFR 19.11(a)(3), as applicable; **AND**	❏	
	We have developed and will implement and maintain written procedures for identifying and responding to emergencies involving radioactive material, including: • Lost, stolen, or missing licensed material, • Exposures to personnel and the public in excess of NRC regulatory limits, • Releases of licensed materials in effluents and the sanitary sewer in excess of NRC regulatory limits,	❏	

Item Number and Title	Suggested Response	Yes	Description Attached
	• Excessive radiation levels or radioactive material concentrations in restricted or unrestricted areas, • Radioactive spills and contamination, • Fires, explosions, and other disasters with the potential for the loss of containment of licensed material, and • Routine contacts with local fire departments and local law enforcement agencies that meet the requirements in 10 CFR 20.1101, 10 CFR 20.2201, 20.2202, 20.2203, and 10 CFR 30.50 and other requirements, as applicable.		
10. Radiation Safety Program 10.7 Surveys	We have developed and will implement and maintain written procedures for a survey program that specifies the performance of radiation and contamination level surveys in restricted and unrestricted areas, personnel contamination monitoring, action levels, and the frequencies and records maintenance of those surveys and monitoring that meet the requirements in 10 CFR 30.53, 10 CFR 20.1501, and 10 CFR 20.2103, as applicable.	❑	
10. Radiation Safety Program 10.8 Dosage Measurement Systems	Describe the types of systems (measurement or combination of measurement and calculation) to be used for the measurement of alpha-, beta-, gamma-, and photon-emitting radioactive drugs; **AND** For each dosage measurement system used to measure the amount of radioactivity in alpha-, beta-, gamma, or photon-emitting radioactive drugs, state: "We have developed, and will implement and maintain, a written procedure for the performance of dosage measurement system checks and tests that meets the requirements in 10 CFR 32.72(c)"; **AND** If applicable, include a sample calculation for determining beta-correction factors for dose calibrators with ionization chambers; **OR** If applicable, include a means for ensuring the accuracy of beta-correction factors supplied by the instrument manufacturer or other entity.	❑ ❑ ❑ ❑	❑ ❑ ❑
10. Radiation Safety Program 10.9 Transportation	The applicant's program for transportation will be examined during inspection, but it should not be submitted in a license application.		N/A

Item Number and Title	Suggested Response	Yes	Description Attached
10. Radiation Safety Program **10.10 Minimization of Contamination**	The applicant does not need to provide a response to this item under the following condition: NRC will consider that the criteria have been met if the applicant's responses meet the criteria for the following sections: • Facilities and Equipment, • Radiation Safety Program - Safe Use of Radionuclides and Emergency Procedures, • Radiation Safety Program - Surveys, • Radiation Safety Program - Leak Testing, and • Waste Management.	☐	☐
10. Radiation Safety Program **10.11 Radioactive Drug Labeling for Distribution**	Describe all labels, indicating the colors to be used, that will accompany the products and describe where each label is placed (e.g., on the "transport radiation shield" or the container used to hold the radioactive drug); and agree to affix the required labels to all "transport radiation shields" and each container used to hold the radioactive drugs.	☐	☐
10. Radiation Safety Program **10.12 Radioactive Drug Shielding for Distribution**	For each radioactive drug to be distributed (except for products intended for redistribution without manipulation and in the manufacturer's original shipping package): • Provide the radionuclide and the maximum activity for each type of container (e.g., vial, syringe);	 ☐	 ☐
	• Describe the type and thickness of the "transport radiation shield" provided for each type of container; and	☐	☐
	• Indicate the maximum radiation level to be expected at the surface of each "transport radiation shield" when the radioactive drug container is filled with the maximum activity.	☐	☐
10. Radiation Safety Program **10.13 Leak Tests**	We have developed and will implement and maintain written procedures for leak testing that meet the requirements in 10 CFR 30.53, 10 CFR 20.1501, and 10 CFR 20.2103.	☐	
11. Waste Management (Pharmacy-Generated Radioactive Wastes)	We have developed, and will implement and maintain written procedures for waste management that meet the requirements in 10 CFR 20.2001(a), 10 CFR 20.2003, and 10 CFR 20.2006, 10 CFR 20.2108, 10 CFR 30.51, as applicable.	☐	
11. Waste Management **11.1 Returned Wastes from Customers**	We have developed and will implement and maintain written procedures for customer return of pharmacy-supplied syringes and vials and their contents, to specify that: • Only pharmacy-supplied syringes and vials and their contents may be returned to the pharmacy,	☐	

Item Number and Title	Suggested Response	Yes	Description Attached
	• Instructions will be provided to radiopharmacy customers for the proper preparation and packaging of the radioactive waste for return to the radiopharmacy, and • Instructions will be provided to pharmacy staff for the pick-up, receipt, and disposal of the returned radioactive waste that meet the requirements in 10 CFR 20.2001(a), 10 CFR 30.33, and 10 CFR 71.5, as applicable.		

APPENDIX E

Sample License

SAMPLE RADIOPHARMACY MATERIALS LICENSE*

1. Radiopharmacy, Inc.
2. 1234 Main Street
   Washington, D.C. 20001

3. License number
4. Expiration date
5. Docket No.
   Reference No.

6. Byproduct, source, and/or special nuclear material	7. Chemical and/or physical form	8. Maximum amount that licensee may possess at any one time under this license
A. Any byproduct material with atomic numbers 1 through 83, except molybdenum-99, technetium-99m, iodine-131 and xenon-133	A. Any	A. 100 millicuries per radionuclide and 1 curie total
B. Molybdenum-99	B. Any	B. 25 curies
C. Technetium-99m	C. Any	C. 25 curies
D. Iodine-131	D. Any	D. 500 millicuries
E. Xenon-133	E. Any	E. 1 curie
F. Fluorine-18	F. Any	F. 20 curies
G. Nitrogen-13	G. Any	G. 5 curies
H. Carbon-11	H. Any	H. 2 curies
I. Oxygen-15	I. Any	I. 2 curies
J. Strontium-82	J. Any	J. 400 millicuries
K. Rubidium-82	K. Any	K. 400 millicuries
L. Strontium-85	L. Any	L. 1 curie
M. Any byproduct material in a brachytherapy source as listed in 10 CFR 35.400	M. Sealed sources	M. 500 millicuries
N. Any byproduct material in a sealed source for diagnosis listed in 10 CFR 35.500	N. Sealed sources	N. 1.5 curies per source and 5.5 curies total
O. Any byproduct material listed in 10 CFR 31.11(a)	O. Prepackaged units for *in vitro* diagnostic tests	O. 10 millicuries
P. Any byproduct material authorized under 10 CFR 35.65	P. Sealed sources	P. 200 millicuries
Q. Depleted Uranium	Q. Metal	Q. 600 kilograms

* Note: Certain information about quantities and locations of radioactive material are no longer released to the public. See Section 5.2.

SAMPLE RADIOPHARMACY MATERIALS LICENSE (cont.)

9. Authorized use:

A. through L.   Preparation and distribution of radioactive drugs, including compounding of iodine-131 and redistribution of used and unused molybdenum-99/technetium-99m and rubidium/strontium-82 generators to authorized recipients in accordance with 10 CFR 32.72. Preparation and distribution of radioactive drugs and radiochemicals including compounding of iodine-131 and redistribution of used and unused molybdenum 99/technetium 99m and rubidium/strontium-82 generators to authorized recipients for nonmedical use.

M. and N.   Redistribution of sealed sources initially distributed by a manufacturer licensed pursuant to 10 CFR 32.74. Redistribution of sealed sources that have been registered either with NRC under 10 CFR 32.210 or with an Agreement State and have been distributed in accordance with an NRC or Agreement State specific license authorizing distribution to persons specifically authorized by an NRC or Agreement State license to receive, possess, and use the devices.

O.   Redistribution to specific licensees or general licensees pursuant to 10 CFR 31.11 provided the packaging and labeling remain unchanged.

P.   Calibration and checking of the licensee's instruments. Redistribution of sealed sources initially distributed by a manufacturer licensed pursuant to 10 CFR 32.74 to authorized recipients and to authorized recipients for nonmedical use.

Q.   Shielding for molybdenum-99/technetium-99m generators.

CONDITIONS

10.   Licensed material may be used only at the licensee's facilities at (fill in the street address of the facility).

11.   Licensed material shall be used by, or under the supervision of:

    A.   A pharmacist working or designated as an authorized nuclear pharmacist in accordance with 10 CFR 32.72(b)(2)(i) or (4).

    B.   Authorized Nuclear Pharmacist(s): [insert names of ANPs].

12.   The Radiation Safety Officer for this license is [insert name of RSO].

13.   In addition to the possession limits in Item 8, the licensee shall further restrict the possession of licensed material to quantities below the minimum limit specified in 10 CFR 30.35(d), 40.36(b), and 70.25(d) for establishing financial assurance for decommissioning.

14.   This license does not authorize distribution to persons exempt from licensing.

15.   A. Sealed sources shall be tested for leakage and/or contamination at intervals not to exceed the intervals specified in the certificate of registration issued by NRC under 10 CFR 32.210 or by an Agreement State.

    B. In the absence of a certificate from a transferor indicating that a leak test has been made within the intervals specified in the certificate of registration issued by NRC

SAMPLE RADIOPHARMACY MATERIALS LICENSE (cont.)

under 10 CFR 32.210 or by an Agreement State prior to the transfer, a sealed source or detector cell received from another person shall not be put into use until tested.

C. Sealed sources need not be tested if they are in storage and are not being used. However, when they are removed from storage for use or transferred to another person, and have not been tested within the required leak test interval, they shall be tested before use or transfer. No sealed source shall be stored for a period of more than 10 years without being tested for leakage and/or contamination.

D. The leak test shall be capable of detecting the presence of 0.005 microcurie (185 Becquerels) of radioactive material on the test sample. If the test reveals the presence of 0.005 microcurie (185 Becquerels) or more of removable contamination, a report shall be filed with the U.S. Nuclear Regulatory Commission in accordance with 10 CFR 30.50(c)(2), and the source shall be removed immediately from service and decontaminated, repaired, or disposed of in accordance with Commission regulations. The report shall be filed within 5 days of the date the leak test result is known with the appropriate NRC Regional Office referenced in Appendix D of 10 CFR Part 20. The report shall specify the source involved, the test results, and corrective action taken.

E. Tests for leakage and/or contamination shall be performed by the licensee or other persons specifically licensed by the Commission or an Agreement State to perform such services. In addition, the licensee is authorized to collect leak test samples for analysis by persons specifically licensed by the Commission or an Agreement State to perform such services. (This condition is used for licensees authorized to collect AND analyze leak test samples.)

16. Sealed sources containing licensed material shall not be opened or sources removed from source holders by the licensee, except as specifically authorized.

17. The licensee shall conduct a physical inventory every six months, or at other intervals approved by NRC, to account for all sealed sources and/or devices received and possessed under the license.

18. Except for maintaining labeling as required by 10 CFR Part 20 or 71, the licensee shall obtain authorization from NRC before making any changes in the sealed source, device, or source-device combination that would alter the description or specifications as indicated in the respective Registration Certificates issued either by the Commission pursuant to 10 CFR 32.210 or by an Agreement State.

19. The licensee is authorized to hold byproduct material with a physical half-life of less than or equal to 120 days for decay-in-storage before disposal without regard to its radioactivity if the licensee:

A. Monitors byproduct material at the surface before disposal and determines that its radioactivity cannot be distinguished from the background radiation level with an appropriate radiation detection survey meter set on its most sensitive scale and with no interposed shielding;

B. Removes or obliterates all radiation labels, except for radiation labels on materials that are within containers and that will be managed as biomedical waste after they have been released from the licensee;

SAMPLE RADIOPHARMACY MATERIALS LICENSE (cont.)

C. Maintains records of the disposal of licensed materials for 3 years. The record must include the date of the disposal, the survey instrument used, the background radiation level, the radiation level measured at the surface of each waste container, and the name of the individual who performed the disposal.

20. The licensee is authorized to retrieve, receive, and dispose of radioactive waste from its customers limited to radiopharmacy-supplied syringes and vials and their contents.

21. The licensee is authorized to transport licensed material only in accordance with the provisions of 10 CFR Part 71, "Packaging and Transportation of Radioactive Material."

22. Except as specifically provided otherwise in this license, the licensee shall conduct its program in accordance with the statements, representations, and procedures contained in the documents, including any enclosures, listed below. The Nuclear Regulatory Commission's regulations shall govern unless the statements, representations, and procedures in the licensee's application and correspondence are more restrictive than the regulations.

A. Application dated [insert date]

B. Letter dated [insert date]

U.S. Nuclear Regulatory Commission

Date: [insert license issue dated]   By:_____[Signature]_____

APPENDIX F

Information Needed for Transfer of Control Application

Information Needed for Transfer of Control Application

Licensees must provide full information and obtain NRC's *prior written consent* before transferring control of the license; some licensees refer to this as "transferring the license." Provide the following information concerning changes of control by the applicant (transferor and/or transferee, as appropriate). If any items are not applicable, so state.

1. The new name of the licensed organization. If there is no change, the licensee should so state.

2. The new licensee contact and telephone number(s) to facilitate communications.

3. Any changes in personnel having control over licensed activities (e.g., officers of a corporation) and any changes in personnel named in the license such as Radiation Safety Officer, Authorized Users, or any other persons identified in previous license applications as responsible for radiation safety or use of licensed material. The licensee should include information concerning the qualifications, training, and responsibilities of new individuals.

4. An indication of whether the transferor will remain in nonlicensed business without the license.

5. A complete, clear description of the transaction, including any transfer of stocks or assets, mergers, etc., so that legal counsel is able, when necessary, to differentiate between name changes and transferring control.

6. A complete description of any planned changes in organization, location, facility, equipment, or procedures (i.e., changes in operating or emergency procedures).

7. A detailed description of any changes in the use, possession, location, or storage of the licensed materials.

8. Any changes in organization, location, facilities, equipment, procedures, or personnel that would require a license amendment even without transferring control.

9. An indication of whether all surveillance items and records (e.g., calibrations, leak tests, surveys, inventories, and accountability requirements) will be current at the time of transfer. Provide a description of the status of all surveillance requirements and records.

10. Confirmation that all records concerning the safe and effective decommissioning of the facility, pursuant to 10 CFR 30.35(g), 40.36(f), 70.25(g), and 72.30(d); public dose; and waste disposal by release to sewers, incineration, radioactive material spills, and on-site burials, have been transferred to the new licensee, if licensed activities will continue at the same location, or to NRC for license terminations.

11. A description of the status of the facility; specifically, the presence or absence of contamination should be documented. If contamination is present, will decontamination

occur before transfer? If not, does the successor company agree to assume full liability for the decontamination of the facility or site?

12. A description of any decontamination plans, including financial assurance arrangements of the transferee, as specified in 10 CFR 30.35, 40.36, and 70.25. Include information about how the transferee and transferor propose to divide the transferor's assets, and responsibility for any cleanup needed at the time of transfer.

13. Confirmation that the transferee agrees to abide by all commitments and representations previously made to NRC by the transferor. These include, but are not limited to: maintaining decommissioning records required by 10 CFR 30.35(g), implementing decontamination activities and decommissioning of the site, and completing corrective actions for open inspection items and enforcement actions.

    With regard to contamination of facilities and equipment, the transferee should confirm, in writing, that it accepts full liability for the site and should provide evidence of adequate resources to fund decommissioning; or the transferor should provide a commitment to decontaminate the facility before transferring control.

    With regard to open inspection items, etc., the transferee should confirm, in writing, that it accepts full responsibility for open inspection items and/or any resulting enforcement actions; or the transferee proposes alternative measures for meeting the requirements; or the transferor provides a commitment to close out all such actions with NRC before license transfer.

14. Documentation that the transferor and transferee agree to transferring control of the licensed material and activity, and the conditions of transfer; and that the transferee is made aware of all open inspection items and its responsibility for possible resulting enforcement actions.

15. A commitment by the transferee to abide by all constraints, conditions, requirements, representations, and commitments identified in the existing license. If not, the transferee must provide a description of its program, to ensure compliance with the license and regulations.

APPENDIX G

Formats for Documenting Training and Experience for Individuals Responsible for Radiation Protection Program

Table G-1   Authorized User or Radiation Safety Officer Training in Basic Radionuclide Handling Techniques

Name (Last, First, Initial)

Location of Training	Dates	Title	Total Hours	Breakdown of Course in Clock Hours				
				RPP	BH	IR	INST	REG
TOTALS								

RPP - Radiation Protection Principles

IR - Ionizing Radiation Units & Characteristics

REG - NRC Regulations and Standards

BH - Biological Hazards

INST - Radiation Detection Instrumentation

Table G-2   Authorized User and Radiation Safety Officer Experience in Handling Radionuclides

(Actual use of radionuclides under the supervision of an Authorized User or Radiation Safety Officer, respectively)

Name (Last, First, Initial)

Isotope(s) Used	Maximum Amount Used at Any One Time	Location of Use	Description of Experience*	Total Hours of Experience

*Description of experience

1.   Shipping, receiving, and performing related radiation surveys.
2.   Using and performing checks for proper operation of dose calibrators, survey meters, and other instruments used to measure photon- and high-energy beta-emitting radionuclides.
3.   Using and performing checks for proper operation of instruments used to measure alpha- and low-energy beta- emitting radionuclides.
4.   Calculating, assaying, and safely preparing radioactive materials.
5.   Use of procedures to prevent or minimize contamination and/or use of proper decontamination procedures.

Documentation of Training and Experience to Identify an Individual on a License as an Authorized Nuclear Pharmacist

I. Experienced Authorized Nuclear Pharmacists

An applicant or licensee that wants to add an experienced Authorized Nuclear Pharmacist (ANP) to its commercial radiopharmacy application or license only needs to provide evidence that the individual is listed as an ANP on a license issued by the NRC or Agreement State, a permit issued by an NRC master materials licensee, a permit issued by an NRC or Agreement State broad-scope licensee, or a permit issued by an NRC master materials broad-scope permittee, and that the individual meets the recentness of training criteria described in 10 CFR 35.59. The applicant also may provide evidence that the individual is identified as an ANP by a commercial nuclear pharmacy authorized to identify ANPs. For individuals who have been previously authorized by, but not listed on, the commercial nuclear pharmacy license, medical broad-scope license, or Master Materials License medical broad-scope permit, the applicant should submit either verification of previous authorizations granted or evidence of acceptable training and experience.

II. Experienced Nuclear Pharmacists Who Only Used Accelerator-Produced Nuclear Materials, or Discrete Sources of Radium-226, or Both, for Nuclear Pharmacy Uses

During the implementation of the EPAct, NRC "grandfathered" nuclear pharmacists that used only accelerator-produced radioactive materials, discrete sources of radium-226 (Ra-226), or both, for nuclear pharmacy uses under the NRC waiver of August 31, 2005, when using these materials for the same uses. Nuclear pharmacists that used accelerator-produced radionuclides or discrete sources of Ra-226 during the effective period of the waiver do not have to meet the requirements of 10 CFR 35.59, or the training and experience requirements in 10 CFR 32.72(b)(2)(i) or (ii), for those materials and uses.

The applicant or licensee that is designating one of these experienced individuals as an ANP under the provisions of 10 CFR 32.72(b)(2)(iii) should document that the individual used only accelerator-produced radionuclides, or discrete sources of Ra-226, for nuclear pharmacy uses during the effective period of the waiver and that the materials were used for the same uses requested. This documentation may be, but is not restricted to, evidence that the individual was listed on an Agreement State or non-Agreement State license or permit authorizing these materials for the requested uses.

III. Applications that Include Individuals for Authorized Nuclear Pharmacist Recognition by NRC

Applicants should submit NRC Form 313A (ANP) to show that the individual meets the correct training and experience criteria in 10 CFR Part 35, Subpart B. There are two primary training and experience routes to qualify an individual as an ANP. The first is by means of certification by a board recognized by NRC and listed on the NRC website (http://www.nrc.gov/materials/miau/med-use-toolkit.html) as provided in 10 CFR 35.55(a). Preceptor attestations must also be submitted for all individuals. The second route is by meeting the structured educational program, supervised work experience, and preceptor attestation requirements in 10 CFR Part 35.55(b).

IV. Recentness of Training

The required training and experience, including board certification, described in 10 CFR Part 35 must be obtained within the 7 years preceding the date of the application, or the individual must document having had related continuing education, retraining, and experience since obtaining the required training and experience. Examples of acceptable continuing education and experience include the following:

- Successful completion of classroom and laboratory review courses that include radiation safety practices relative to the practice of nuclear pharmacy, and

- Practical experience in nuclear pharmacy under the supervision of an ANP at the same or another licensed facility that is authorized as a nuclear pharmacy.

V. General Instructions and Guidance for Filling Out NRC Form 313A Series

If the applicant wishes to identify a license and it is an Agreement State license, the applicant should provide a copy of the license. If the applicant wishes to identify a Master Materials License permit, the applicant should provide a copy of the permit. If the applicant wishes to identify a preceptor who is authorized under a broad-scope license or broad-scope permit of a Master Materials License, the applicant should provide a copy of the permit issued by the broad-scope licensee/permittee. Alternatively, the applicant may provide a statement signed by the Radiation Safety Officer or chairperson of the Radiation Safety Committee similar to the following: "_____(name of preceptor) is authorized under _____(name of licensee/permittee) broad-scope license number_____ to be an ANP during _____ (time frame)."

INTRODUCTORY INFORMATION

Name of Individual

Provide the individual's complete name so that NRC can distinguish the training and experience received from that received by others with a similar name.

Note: Do not include personal or private information (e.g., date of birth, Social Security number, home address, personal phone number) as part of your qualification documentation.

State or Territory where Licensed

Note that the NRC requires pharmacists to be licensed by a State or territory of the United States, the District of Columbia, or the Commonwealth of Puerto Rico to practice pharmacy.

Requested Authorization(s)

Check all authorizations that apply and fill in the blanks as provided.

Part I. Training and Experience

There are always multiple pathways provided for each training and experience section. Select the applicable one.

Item 1. Board Certification

The applicant or licensee may use this pathway if the proposed nuclear pharmacist is certified by a board recognized by NRC (to confirm that NRC recognizes that board's certifications, see NRC's web page http://www.nrc.gov/materials/miau/med-use-toolkit.html.

Notes:

- An individual that is board-eligible will not be considered for this pathway until the individual is actually board-certified. Further, individuals holding other board certifications will also not be considered for this pathway.

- The applicant or licensee must provide a copy of the board certification and completed attestation as indicated on the attached NRC Form 313A (ANP).

- As indicated on the form, additional information is needed if the board certification was obtained more than 7 years ago.

Item 2. Structured Educational Program for a Proposed Authorized Nuclear Pharmacist

This pathway is used for those individuals not listed on the license as an ANP, who do not meet the requirements for the board certification pathway.

The regulatory requirements refer to a structured educational program consisting of both (a) classroom and laboratory training, and (b) supervised practical experience in nuclear pharmacy. All hours credited to classroom and laboratory training must relate directly to radiation safety and safe handling of byproduct material and be allocated to one of the topics in 10 CFR 35.55 (b)(1)(i).

The proposed ANP may receive the required classroom and laboratory training, and supervised practical experience at a single training facility or at multiple training facilities; therefore, space is provided to identify each location and date of training or experience. The date should be provided in the month/day/year (mm/dd/yyyy) format. Under the "classroom and laboratory training," provide the number of clock hours spent on each of the topics listed in the regulatory requirements.

The proposed ANP may obtain the required "classroom and laboratory training" in any number of settings, locations, and educational situations. For example, at some medical teaching/university institutions, a course may be provided for that particular need and taught on consecutive days. In other training programs, the period may be a semester or quarter as part of the formal curriculum. Also, the classroom and laboratory training may be obtained using a variety of other instructional methods. Therefore, NRC will broadly interpret "classroom and laboratory training" to include various types of instruction, including online training, as long as it meets the specific clock hour requirements and the subject matter relates to radiation safety and safe handling of byproduct material for the uses requested.

Under the "supervised practical experience" section of the form, provide the number of clock hours for each topic. The supervised practical experience topics for the nuclear pharmacists include all the basic elements in the practice of nuclear pharmacy. Therefore, all the hours of supervised experience are allocated to these topics.

Note: As indicated on the form, additional information is needed if the training and/or supervised practical experience was completed more than 7 years ago.

Part II. Preceptor Attestation

The NRC defines the term "preceptor" in 10 CFR 35.2, "Definitions," to mean "an individual who provides, directs, or verifies training and experience required for an individual to become an authorized user, an authorized medical physicist, an authorized nuclear pharmacist, or a Radiation Safety Officer." While the supervising individual for the practical experience in nuclear pharmacy may also be the preceptor, the preceptor does not have to be the supervising individual as long as the preceptor directs or verifies the training and experience required. The preceptor must attest in writing regarding the training and experience of any individual to serve as an authorized individual and attest that the individual has satisfactorily completed the appropriate training and experience criteria and has achieved a level of competency sufficient to function independently. This preceptor also has to meet specific requirements.

The NRC Form 313A (ANP) Part II - Preceptor Attestation has two sections. The preceptor must select either the board certification or the structured educational program when filling out the first section on this page. The second and final sections of the page request specific information about the preceptor's authorization to use licensed material in addition to the preceptor's signature.

NRC FORM 313A (ANP) (10-2006)	U.S. NUCLEAR REGULATORY COMMISSION	
AUTHORIZED NUCLEAR PHARMACIST TRAINING AND EXPERIENCE AND PRECEPTOR ATTESTATION **[10 CFR 35.55]**		APPROVED BY OMB: NO. 3150-0120 EXPIRES: 10/31/2008

Name of Proposed Authorized Nuclear Pharmacist	State or Territory Where Licensed

PART I — TRAINING AND EXPERIENCE
(Select one of the two methods below)

* Training and Experience, including board certification, must have been obtained within the 7 years preceding the date of application or the individual must have obtained related continuing education and experience since the required training and experience was completed. Provide dates, duration, and description of continuing education and experience related to the nuclear pharmacy uses.

☐ 1. **Board Certification**

    a. Provide a copy of the board certification.

    b. Skip to and complete Part II Preceptor Attestation.

☐ 2. **Structured Educational Program for Proposed Authorized Nuclear Pharmacist**

    a. Classroom and Laboratory Training:

Description of Training	Location of Training	Clock Hours	Dates of Training*
Radiation physics and instrumentation			
Radiation protection			
Mathematics pertaining to the use and measurement of radioactivity			
Chemistry of byproduct material for medical use			
Radiation biology			
Total Hours of Training:			

APPENDIX G

NRC FORM 313A (ANP) (10-2006)			U.S. NUCLEAR REGULATORY COMMISSION

AUTHORIZED NUCLEAR PHARMACIST TRAINING AND EXPERIENCE AND PRECEPTOR ATTESTATION (continued)

2. Structured Educational Program for Proposed Authorized Nuclear Pharmacist (continued)

   b. Supervised Practical Experience in a Nuclear Pharmacy.

Description of Experience	Location of Experience/License or Permit Number of Facility	Clock Hours	Dates of Experience*
Shipping, receiving, and performing related radiation surveys			
Using and performing checks for proper operation of instruments used to determine the activity of dosages, survey meters, and, if appropriate, instruments used to measure alpha- or beta-emitting radionuclides			
Calculating, assaying, and safely preparing dosages for patients or human research subjects			
Using administrative controls to avoid medical events in administration of byproduct material			
Using procedures to prevent or minimize radioactive contamination and using proper decontamination procedures			
Total Hours of Experience:			
Supervising Individual			

   c. Go to and complete Part II Preceptor Attestation.

PAGE 2

NRC FORM 313A (ANP)
(10-2006)

U.S. NUCLEAR REGULATORY COMMISSION

AUTHORIZED NUCLEAR PHARMACIST TRAINING AND EXPERIENCE AND PRECEPTOR ATTESTATION (continued)

PART II – PRECEPTOR ATTESTATION

Note: This part must be completed by the individual's preceptor. The preceptor does not have to be the supervising individual as long as the preceptor provides, directs, or verifies training and experience required. If more than one preceptor is necessary to document experience, obtain a separate preceptor statement from each.

First Section
Check one of the following:

Board Certification

☐ I attest that _____ has satisfactorily completed the requirements in
Name of Proposed Authorized Nuclear Pharmacist

10 CFR 35.55(a)(1), (a)(2), and (a)(3) and has achieved a level of competency sufficient to function independently as an authorized nuclear pharmacist.

OR

Structured Educational Program

☐ I attest that _____ has satisfactorily completed a 700-hour structured
Name of Proposed Authorized Nuclear Pharmacist

educational program consisting of both 200 hours of classroom and laboratory training, and practical experience in nuclear pharmacy, as required by 10 CFR 35.55(b)(1) and has achieved a level of competency sufficient to function independently as an authorized nuclear pharmacist.

- -

Second Section
Complete the following for preceptor attestation and signature:

I am an Authorized Nuclear Pharmacist for _____
Nuclear Pharmacy or Medical Facility

_____
License/Permit Number

Name of Preceptor	Signature	Telephone Number	Date

PAGE 3

APPENDIX H

Typical Duties and Responsibilities of the Radiation Safety Officer

Typical Duties and Responsibilities of the Radiation Safety Officer

The RSO's duties and responsibilities include ensuring radiological safety and compliance with NRC and DOT regulations, and with the conditions of the license (see Figure H.1). Typically, these duties and responsibilities include ensuring that:

- General surveillance is provided over all activities involving radioactive material, including routine monitoring, special surveys, and responding to events.

- Incidents are responded to, investigated, their cause(s) and appropriate corrective action(s) are identified, and timely corrective action(s) are taken.

- Proper authorities are notified of incidents such as damage, fire, or theft.

- Corrective actions are developed, implemented, and documented when violations of regulations or license conditions or program weaknesses are identified.

- All activities are immediately terminated following any unsafe condition or activity that is found to be a threat to public health and safety.

- He or she is the primary source of radiation protection information for personnel at all levels of responsibility.

- All radiation workers are properly trained.

- Procedures for the safe use of radioactive materials are developed and implemented.

- The licensee's procedures and controls, based upon sound radiation protection principles, are periodically reviewed to ensure that occupational doses and doses to members of the public are as low as is reasonably achievable (ALARA). Documentation is maintained to demonstrate, by measurement or calculation, that the total effective dose equivalent to the individual member of the public likely to receive the highest dose from the licensed operation does not exceed the annual limit.

- Prospective evaluations are performed of occupational exposures, and those individuals likely to receive, in one year, a radiation dose in excess of 10% of the allowable limits are provided personnel monitoring devices.

- When necessary, personnel monitoring devices are used and exchanged at the proper intervals, and records of the results of such monitoring are maintained.

- The performance of fume hoods and gloveboxes used for volatile radioactive material work are monitored for proper operation.

- The receipt, opening, and delivery of all packages of radioactive material arriving at the nuclear pharmacy are overseen and coordinated.

- An inventory of all radioactive materials is maintained and the types and quantities of radionuclides at the facility are limited to the forms and amounts authorized by the license.

- Sealed sources are leak-tested at required intervals.

- There is effective management of the radioactive waste program, including effluent monitoring.

- Packaging and transport of radioactive material is in accordance with all applicable DOT requirements.

- An up-to-date license is maintained and amendment and renewal requests and notifications of new ANPs are submitted in a timely manner.

- Radiation Safety Program audits are performed at least annually and documented.

- He or she acts as liaison to NRC.

- All required records are properly maintained.

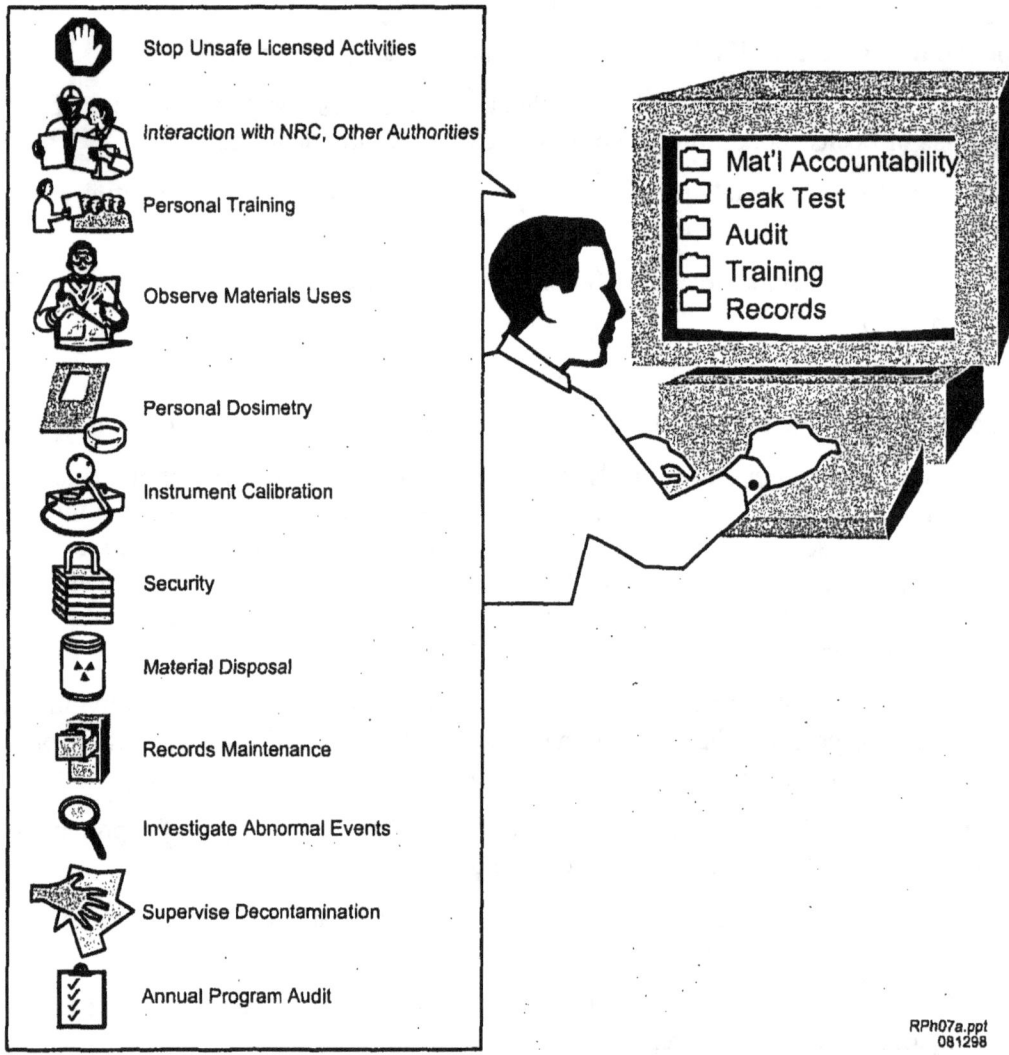

Figure H.1 Typical Duties and Responsibilities of the RSO.

APPENDIX I

Suggested Commercial Radiopharmacy
Audit Checklist

Suggested Commercial Radiopharmacy Audit Checklist

Note: All areas indicated in audit notes may not be applicable to every license and may not need to be addressed during each audit. For example, licensees do not need to address areas that do not apply to the licensee's activities, and activities which have not occurred since the last audit need not be reviewed at the next audit.

Date of This Audit _____   Date of Last Audit_____

Next Audit Date _____

Auditor _____   Date _____

_____
(Signature)

Management Review_____   Date_____

_____
(Signature)

Audit History

A. Last audit of this location conducted on (date)

B. Were previous audits conducted at intervals not to exceed 12 months? [10 CFR 20.1101]

C. Were records of previous audits maintained? [10 CFR 20.2102]

D. Were any deficiencies identified during last two audits or two years, whichever is longer?

E. Were corrective actions taken? (Look for repeated deficiencies.)

Organization and Scope of Program

A. If the mailing address or places of use changed, was the license amended? [10 CFR 30.34]

B. If ownership changed or bankruptcy filed, was NRC's prior consent obtained or was NRC notified? [10 CFR 30.34]

C. Authorized Nuclear Pharmacists

   1. New ANP since last audit? If so, does new ANP meet NRC training requirements? [10 CFR 32.72, 10 CFR 35.2, 10 CFR 35.55(b)]

   2. If an individual began work as an ANP, was NRC notified within 30 days or was license amended? [10 CFR 32.72]

D. Radiation Safety Officer

   1. New RSO since last audit? If so, does new RSO meet NRC training requirements?

   2. If the RSO was changed, was license amended?

   3. Is RSO fulfilling his/her duties?

    4. To whom does RSO report?

E. Authorized Users

    1. New AU since last audit? If so, does new AU meet NRC training requirements?

    2. If an AU was added, was license amended?

F. If the designated contact person for NRC changed, was NRC notified?

G. Type and quantity of byproduct material

    1. Does the license authorize all of NRC-regulated radionuclides possessed?

    2. Is actual possession of those radionuclides within the limits on the license?

Facilities

A. Are facilities as described in NRC license application?

B. If facilities have changed, has NRC license been amended?

Equipment and Instrumentation

A. Are there sufficient portable and fixed radiation monitors?

B. Do survey meters meet NRC's criteria? [10 CFR 20.1501]

C. Are calibration records maintained? [10 CFR 20.2103(a)]

D. Is there sufficient shielding (L-block, etc.) for work with radionuclides?

E. Are generators housed in separate room and/or properly shielded to keep doses ALARA?

F. Are procedures established for identifying, evaluating, and reporting safety component defects? [10 CFR 21.21]

G. Dose Calibrators for Photon-emitters [10 CFR 32.72©]

    1. Constancy, at least once each day prior to assay of patient dosages (±10%)?

    2. Linearity, at installation and at required frequency (±10%)?

    3. Geometry dependence, at installation (±10%)?

    4. Accuracy, at installation and at required frequency (±10%)?

    5. After repair, adjustment, or relocation of the dose calibrator, were appropriate tests above repeated?

H. Dose Measurement Systems for Beta- and Alpha-emitters [10 CFR 32.72©]

    1. Calibrated for each isotope used, with that isotope?

    2. Constancy, at least once each day prior to assay of patient dosages (±10%)?

    3. Geometry dependence, at installation (±10%)?

    4. Accuracy, at installation and at manufacturer's recommended frequency (±10%)?

5. Linearity, at installation and at manufacturer's recommended frequency (±10%)?

6. After repair, adjustment, or relocation of the dose calibrator, were appropriate tests above repeated?

Area Surveys and Contamination Control [10 CFR 20.1501]

A. Are area surveys being performed at applicable locations and required frequencies? Are records maintained? [10 CFR 20.2103]

B. Are removable contamination surveys being performed at applicable locations and required frequencies? Are records maintained? [10 CFR 20.2103]

C. Is appropriate corrective action taken and documented when excess radiation or contamination levels are detected?

Leak Tests

A. Was each sealed source leak tested every six months or at other prescribed intervals?

B. Was the leak test performed according to the license?

C. Are records of results retained with the appropriate information included?

D. Were any sources found leaking and if yes, was NRC notified?

Sealed Source Inventory

A. Is a record kept showing the receipt of each sealed source? [10 CFR 30.51(a)(1)]

B. Are all sealed sources physically inventoried every six months?

C. Are records of inventory results with appropriate information maintained?

Training and Instructions to Workers

A. Were all workers who are likely to exceed 1 mSv (100 mrem) in a year instructed per [10 CFR 19.12]? Was refresher training provided, as needed? [10 CFR 19.12] Are records maintained?

B. Were other workers trained as needed (e.g., radiopharmacy technicians, authorized users, couriers/drivers, ancillary personnel)? [10 CFR 30.33] Are records maintained?

C. Are workers knowledgeable of applicable 10 CFR Part 20 radiation protection procedures, emergency response procedures, and license conditions?

D. Was HAZMAT training provided, if required? [49 CFR 172.700, 49 CFR 172.701, 49 CFR 172.702, 49 CFR 172.704]

Material Use Control and Transfer

A. Are restricted and unrestricted areas delineated?

B. Are radioactive materials that are stored in a controlled or unrestricted area secured from unauthorized access or removal? [10 CFR 20.1801]

C. Are radioactive materials that are in a controlled or unrestricted area and not in storage controlled and maintained under constant surveillance? [10 CFR 20.1802]

D. Are there procedures for receiving and opening packages? [10 CFR 20.1906]

E. Is byproduct material transferred only to authorized recipients? [10 CFR 30.41, 10 CFR 32.71, 10 CFR 32.72, 10 CFR 32.74]

F. Are records kept of receipt and transfer? [10 CFR 30.51]

Personnel Radiation Protection

A. Are ALARA considerations incorporated into the Radiation Protection Program? [10 CFR 20.1101(b)]

B. Were prospective evaluations performed showing that unmonitored individuals receive ≤10% of limit? [10 CFR 20.1502(a)]

C. Did unmonitored individuals' activities change during the year which could put them over 10% of limit?

D. If yes to C. above, was a new evaluation performed?

E. Is external dosimetry required (individuals likely to receive >10% of limit)? And is dosimetry provided to these individuals?

    1. Is the dosimetry supplier NVLAP-approved? [10 CFR 20.1501©]

    2. Are the dosimeters exchanged at appropriate frequency?

    3. Are dosimetry reports reviewed by the RSO when they are received?

    4. Are the records on NRC Forms or equivalent? [10 CFR 20.2104(d), 10 CFR 20.2106©]

        a. NRC-Form 4 "Cumulative Occupational Exposure History" completed?

        b. NRC-Form 5 "Occupational Exposure Record for a Monitoring Period" completed?

    5. Declared pregnant worker/embryo/fetus

        a. If a worker declared her pregnancy, did licensee comply with [10 CFR 20.1208]?

        b. Were records kept of embryo/fetus dose per [10 CFR 20.2106(e)]?

F. Are individuals monitored for internal dose if they are likely to receive >10% of ALI?

G. Are workers notified annually of their exposures?

H. Are records of exposures, surveys, monitoring, and evaluations maintained? [10 CFR 20.2102, 10 CFR 20.2103, 0 CFR 20.2106]

Waste Management

A. Waste storage areas

   1. Is storage area properly posted? [10 CFR 20.1902]

   2. Are containers properly labeled? [10 CFR 20.1904]

B. Decay-in-Storage

   1. Do radionuclides being stored all have half-lives less than 120 days?

   2. Are radionuclides being segregated for storage according to half-life?

   3. Before waste is disposed of:

      a. Is a survey performed at the container surface with an appropriate survey instrument set on its most sensitive scale, with no interposed shielding, to determine that its radioactivity cannot be distinguished from background?

      b. Are all radiation labels removed or obliterated, as appropriate?

   4. Are records kept?

C. Disposal by release into sanitary sewerage.

   1. Is licensed material readily soluble (or readily dispersible biological material) in water? [IN 94-07]: Solubility Criteria for Liquid Effluent Releases to Sanitary Sewerage Under the Revised 10 CFR Part 20. [10 CFR 20.2003]

   2. Does the quantity of licensed material that the licensee releases into the sewer each month averaged over the monthly volume of water released into the sewer not exceed the concentration specified in 10 CFR Part 20, Appendix B, Table 3?

   3. If more than one radionuclide is released, does the sum of the ratios of the average monthly discharge of a radionuclide to the corresponding limit in 10 CFR Part 20, Appendix B, Table 3 not exceed unity?

   4. Does the total quantity of licensed material released into the sanitary sewerage system in a year not exceed the limits specified in 10 CFR 20.2003(a)(4)?

D. Transfer to Authorized Recipient

   1. Is waste being transferred to a person specifically authorized to receive it? [10 CFR 20.2001]

   2. Is waste properly manifested? [10 CFR 20.2006]

Receipt of Radioactive Waste from Customers

A. Does returned waste consist only of items that contained radioactive materials that the radiopharmacy supplied (e.g., pharmacy supplied syringes, vials)?

B. Are waste packages checked for removable contamination upon receipt?

Effluents

A. Are effluents from materials being maintained ALARA?

B. Are fume hoods checked to confirm an adequate airflow?

C. Is effluent monitored to determine activity being released?

D. Are filters being maintained according to the manufacturer's instructions and pharmacy procedures?

Public Dose

A. Is public access to radioactive materials and exposure to effluents controlled in a manner to keep doses below 1 mSv (100 mrem) in a year? [10 CFR 20.1301(a)(1)]

B. Are air emissions maintained below constraint limit of 0.1 mSv (10 Millirem) in a year? [10 CFR 20.1101(d)]

C. Are survey or prospective evaluations performed per 10 CFR 20.1501(a)? Have there been any additions or changes to the storage, security, or use of surrounding areas that would necessitate a new survey or evaluation?

D. Do unrestricted area radiation levels exceed 0.02 mSv (2 mrem) in any one hour? [10 CFR 20.1301(a)(2)]

E. Are records maintained? [10 CFR 20.2103, 10 CFR 20.2107]

Use and Emergency Procedures

A. Are procedures for safe use of radioactive materials and emergency procedures developed and implemented?

B. Do the procedures contain the required elements?

C. Are radioactive materials being handled safely?

D. Does the staff wear protective clothing and personnel monitors as appropriate?

E. Is assistance coordinated with outside agencies for emergency response (e.g., fire department)?

F. Did any emergencies occur?

    1. If so, were they handled properly?

    2. Were appropriate corrective actions taken?

    3. Was NRC notification or reporting required? [10 CFR 20.2201, 10 CFR 20.2202, 10 CFR 20.2203]

Transportation

A. Are DOT-7A or other authorized packages used? [49 CFR 173.415]

B. Are package performance test records on file?

C. Does each package have two labels (ex. Yellow-II) with Transportation Index (TI), Nuclide, Activity, and Hazard Class? [49 CFR 172.403, 49 CFR 173.441]

D. Are packages properly marked? [49 CFR 172.301, 49 CFR 172.302, 49 CFR 172.304, 49 CFR 172.310, 49 CFR 172.324]

E. Are packages closed and sealed during transport? [49 CFR 173.412(a), 49 CFR 173.475(f)]

F. Are shipping papers prepared and used? [49 CFR 172.200(a)]

G. Do shipping papers contain proper entries? (Shipping name; Hazard Class; Identification Number (UN Number); Total Quantity; Package Type; Nuclide; Reportable Quantity (RQ); Physical and Chemical Form; Activity (SI units required); category of label; TI; Shipper's Name, Certification, and Signature; Emergency Response Phone Number; Emergency Response Information; and Cargo Aircraft Only (if applicable)) [49 CFR 172.200, 49 CFR 172.201, 49 CFR 172.202, 49 CFR 172.203, 49 CFR 172.204, 49 CFR 172.604]

H. Are shipping papers within driver's reach and readily accessible during transport? [49 CFR 177.817(e)].

I. Are packages secured against movement? [49 CFR 177.834]

J. Are incidents reported to DOT? [49 CFR 171.15, 49 CFR 171.16]

Auditor's Independent Survey Measurements (If Made)

A. Describe the type, location, and results of measurements. Also note the survey instrument used, serial number, and calibration date. Does any radiation level exceed regulatory limits? [10 CFR 20.1501(a), 10 CFR 20.1502(a)]

Notification and Reports

A. Was any radioactive material lost or stolen? Were reports made? [10 CFR 20.2201, 10 CFR 30.50]

B. Did any reportable incidents occur? Were reports made? [10 CFR 20.2202, 10 CFR 30.50]

C. Did any overexposures or high radiation levels occur? Were they reported? [10 CFR 20.2203, 10 CFR 30.50]

D. Were any contaminated packages or packages with surface radiation levels exceeding 200 mrem received? Were they reported to NRC?

E. If any events (as described in items A through D above) did occur, what was the root cause? Were appropriate notifications made and corrective actions taken?

F. Is the management/RSO aware of the telephone number for the NRC Emergency Operations Center? [(301) 816-5100]

Posting and Labeling

A. Is NRC-Form 3, "Notice to Workers" posted? [10 CFR 19.11]

B. Are NRC regulations and license documents posted or is a notice posted? [10 CFR 19.11, 10 CFR 21.6; Section 206 of Energy Reorganization Act of 1974]

C. Are other posting and labeling requirements met? [10 CFR 20.1902, 10 CFR 20.1904]

Recordkeeping for Decommissioning

A. Are records kept of information important to decommissioning? [10 CFR 30.35(g)]

B. Do records include all information outlined in 10 CFR 30.35(g)?

Bulletins and Information Notices

A. Are NRC Bulletins, NRC Information Notices, NMSS Newsletters received?

B. Are appropriate training and action taken in response?

Special License Conditions or Issues

A. Did an auditor review special license conditions or other issues?

Deficiencies Identified in Audit; Corrective Actions

A. Summarize problems/deficiencies identified during the audit.

B. If problems/deficiencies were identified in this audit, describe corrective actions planned or taken by the facility. Include date(s) when corrective actions are implemented.

C. Provide any other recommendations for improvement.

Evaluation of Other Factors

A. Is licensee's senior management appropriately involved with the Radiation Protection Program and/or RSO oversight?

B. Does the RSO have sufficient time to perform his/her radiation safety duties?

C. Does the licensee have sufficient staff to support the Radiation Protection Program?

APPENDIX J

General Radiation Monitoring Instrument Specifications and Model Survey Instrument Calibration Program

General Radiation Monitoring Instrument Specifications and Model Survey Instrument Calibration Program

The specifications in Table J.1 will help applicants and licensees choose the proper radiation detection equipment for monitoring the radiological conditions at their facility(ies).

Table J.1   Typical Survey Instruments[1] - *Instruments used to measure radiological conditions at licensed facilities.*

Portable Instruments Used for Contamination and Ambient Radiation Surveys			
Detectors	**Radiation**	**Energy Range**	**Efficiency**
Exposure Rate Meters	Gamma, X-ray	mR-R	N/A
Count Rate Meters			
GM	Alpha	All energies (dependent on window thickness)	Moderate
	Beta	All energies (dependent on window thickness)	Moderate
	Gamma	All energies	<1%
NaI Scintillator	Gamma	All energies (dependent on crystal thickness	Moderate
Plastic Scintillator	Beta	C-14 or higher (dependent on window thickness)	Moderate
Stationary Instruments Used to Measure Wipe, Bioassay, and Effluent Samples			
Detectors	**Radiation**	**Energy Range**	**Efficiency**
LSC*	Alpha	All energies	High
	Beta	All energies	High
	Gamma		Moderate
Gamma Counter (NaI)*	Gamma	All energies	High
Gas Proportional	Alpha	All energies	High
	Beta	All energies	Moderate
	Gamma	All energies	<1%

[1]   Table from The Health Physics & Radiological Health Handbook, Third Edition, Edited by Bernard Shleien, 1998 (except * items).

Model Instrument Calibration Program

Training

Before allowing an individual to perform survey instrument calibrations, the RSO will ensure that he or she has sufficient training and experience to perform independent survey instrument calibrations.

Classroom training may be in the form of lecture, videotape, or self-study and will cover the following subject areas:

- Principles and practices of radiation protection;

- Radioactivity measurements, monitoring techniques, and using instruments;

- Mathematics and calculations basic to using and measuring radioactivity; and

- Biological effects of radiation.

Appropriate on-the-job-training consists of the following:

- Observing authorized personnel performing survey instrument calibration; and

- Conducting survey meter calibrations under the supervision and in the physical presence of an individual authorized to perform calibrations.

Facilities and Equipment for Calibration of Dose Rate or Exposure Rate Instruments

- To reduce doses received by individuals not calibrating instruments, calibrations will be conducted in an isolated area of the facility or at times when no one else is present;

- Individuals conducting calibrations will wear assigned dosimetry; and

- Individuals conducting calibrations will use a calibrated and operable survey instrument to ensure that unexpected changes in exposure rates are identified and corrected.

Model Procedure for Calibrating Survey Instruments

A radioactive sealed source(s) used for calibrating survey instruments will:

- Approximate a point source,

- Have its apparent source activity or the exposure rate at a given distance traceable by documented measurements to a standard certified to be within ± 5% accuracy by National Institutes of Standards and Technology (NIST),

- Approximate the same energy and type of radiation as the environment in which the calibrated device will be employed, and

- For dose rate and exposure rate instruments, the source should be strong enough to give an exposure rate of at least about 7.7×10^{-6} coulombs/kilogram/hour (30 mR/hr) at 100 cm (e.g., 3.1 gigabecquerels (85 mCi) of cesium-137 or 7.8×10^{2} megabecquerels (21 mCi) of cobalt-60).

The three kinds of scales frequently used on dose or dose rate survey meters are calibrated as follows[2]:

- Linear readout instruments with a single calibration control for all scales shall be adjusted at the point recommended by the manufacturer or at a point within the normal range of use. Instruments with calibration controls for each scale shall be adjusted on each scale. After adjustment, the response of the instrument shall be checked at approximately 20% and 80% of full scale. The instrument's readings shall be within $\pm$ 15% of the conventionally true values for the lower point and $\pm$ 10% for the upper point.

- Logarithmic readout instruments, which commonly have a single readout scale spanning several decades, normally have two or more adjustments. The instrument shall be adjusted for each scale according to site specifications or the manufacturer's specifications. After adjustment, calibration shall be checked at a minimum of one point on each decade. Instrument readings shall have a maximum deviation from the conventionally true value of no more than 10% of the full decade value.

- Meters with a digital display device shall be calibrated the same as meters with a linear scale.

Notes:

- Readings above 2.58×10^{-4} coulomb/kilogram/hour (1 R/hr) need not be calibrated, but such scales should be checked for operation and response to radiation.

- The inverse square and radioactive decay law should be used to correct changes in exposure rate due to changes in distance or source decay.

Surface Contamination Measurement Instruments[2]

- Survey meters' efficiency must be determined by using radiation sources with similar energies and types of radiation that the survey instrument will be used to measure.

- If each scale has a calibration potentiometer, the reading shall be adjusted to read the conventionally true value at approximately 80% of full scale, and the reading at approximately 20% of full scale shall be observed. If only one calibration potentiometer is available, the reading shall be adjusted at mid-scale on one of the scales, and readings on the other scales shall be observed. Readings shall be within 20% of the conventionally true value.

---

[2] ANSI N323A-1997, "Radiation Protection Instrumentation Test and Calibration."

Model Procedures for Calibrating Liquid Scintillation Counters, Gamma Counters, Gas-Flow Proportional Counters, and Multichannel Analyzers

A radioactive sealed source used for calibrating instruments will do the following:

- Approximate the geometry of the samples to be analyzed,

- Have its apparent source activity traceable by documented measurements to a standard certified to be within ±.5% accuracy by NIST, and

- Approximate the same energy and type of radiation as the samples that the calibrated device will be used to measure.

Calibration

- Calibration must produce readings within ± 20% of the actual values over the range of the instrument.

- Calibration of liquid scintillation counters will include quench correction.

Calibration Records

Calibration reports, for all survey instruments, will indicate the procedure used and the data obtained. The description of the calibration will include:

- The owner or user of the instrument;

- A description of the instrument, including the manufacturer's name, model number, serial number, and type of detector;

- A description of the calibration source, including the exposure rate at a specified distance or activity on a specified date;

- For each calibration point, the calculated exposure rate or count rate, the indicated exposure rate or count rate, the deduced correction factor (the calculated exposure rate or count rate divided by the indicated exposure rate or count rate), and the scale selected on the instrument;

- For instruments with external detectors, the angle between the radiation flux field and the detector (i.e., parallel or perpendicular);

- For instruments with internal detectors, the angle between radiation flux field and a specified surface of the instrument;

- For detectors with removable shielding, an indication whether the shielding was in place or removed during the calibration procedure;

- The exposure rate or count rate from a check source, if used; and

- The name of the person who performed the calibration and the date it was performed.

The following information will be attached to the instrument as a calibration sticker or tag:

- For exposure rate meters, the source isotope used to calibrate the instrument (with correction factors) for each scale;

- The efficiency of the instrument, for each isotope the instrument will be used to measure (if efficiency is not calculated before each use);

- For each scale or decade not calibrated, an indication that the scale or decade was checked only for function but not calibrated;

- The date of calibration and the next calibration due date; and

- The apparent exposure rate or count rate from the check source, if used.

Air Sampler Calibration

In order to assess accurately the air concentration of radioactive materials in a given location, the volume of air sampled and the quantity of contaminant in the sample must be determined. Accurate determination of the volume of air sampled requires standard, reproducible, and periodic calibration of the air metering devices that are used with air sampling instruments.

The publication entitled "Air Sampling Instruments" found in the 9th Edition, American Conference of Governmental Industrial Hygienists, 2001, provides guidance on total air sample volume calibration methods acceptable to NRC staff, as supplemented below.

Frequency of Calibration

- A licensee committed to a routine or emergency air sampling program should perform an acceptable calibration of all airflow or volume metering devices at least annually (see Regulatory Guide 8.25).

- Special calibrations should be performed at any time there is reason to believe that the operating characteristics of a metering device have been changed, by repair or alteration, or whenever system performance is observed to have changed significantly.

- Routine instrument maintenance should be performed as recommended by the manufacturer.

- Primary or secondary standard instruments used to calibrate air sampling instruments should be inspected frequently for consistency of performance.

Error Limit For Measurement of Air Sample Volume

Most methods of calibrating airflow or air volume metering devices require direct comparison to a primary or secondary standard instrument, to determine a calibration curve or a correction factor. An example of a primary standard is a spirometer that measures total air volume directly with high precision by liquid displacement. An example of a secondary standard is a wet-test meter that has been calibrated against a primary standard. Primary standards are usually accurate to within ± 1% and secondary standards to within ± 2%.

The following are significant errors associated with determining the total air volume sampled:

E_c: The error in determining the calibration factor. (An acceptable estimate is the percentage error associated with the standard instrument used in the calibration.)[3]

E_s: Intrinsic error in reading the meter scale. (An acceptable estimate is the percentage equivalent of one-half of the smallest scale division, compared to the scale reading.)

E_t: The percentage error in measurement of sampling time that should be kept within 1%.

E_v: The most probable value of the cumulative percentage error in the determination of the total air volume sampled. This can be calculated from the following equation, provided there are no additional significant sources of errors:

$$E_V = [E_s^2 + E_c^2 + E_t^2]^{1/2}$$

The most probable value of the cumulative error E_V, in the determination of total volume, should be less than 20%.

A sample calculation of the most probable value of the cumulative error in total volume measured is as follows: If accuracies of the scale reading, the calibration factor, and sample time are ± 4, 2, and 1%, respectively, and there are no other significant sources of error, the cumulative error would be:

$$E_V = [4^2 + 2^2 + 1^2]^{1/2} = 4.58\% \text{ or approx. } 5\%$$

If there are significant differences in pressure and temperature between the calibration site and the sampling site, appropriate corrections should be made using the ideal gas laws provided below:

$$Vs = V1 * (P1/760) * (273/T1)$$

where   Vs = volume at standard conditions (760 mm & 0C)

          V1 = volume measured at conditions P1 and T1

          T1 = temperature of V1 in K

          P1 = pressure of V1 in mm Hg

---

[3] The calibration factor should be based on two kinds of determinations. First, correction factors should be determined at several flow rates distributed over the full-scale range. Each flow rate correction factor should be determined while adjusting flow rates upscale and again while adjusting flow rates downscale, and the two sets of data should be compared. Second, subsequent calibrations should compare the new correction factors to those determined during the previous calibration. If observed differences are significant compared to the overall volume error limit of 20%, an additional error term should be included in the calculation above.

Documentation of Calibration of Air Metering Devices

The licensee should maintain records of all routine and special calibrations of airflow or volume metering devices, including the primary or secondary standard used, method employed, and estimates of accuracy of the calibrated metering devices. All instruments should be clearly labeled as to the date and results of the most recent calibration and should include the appropriate correction factors to be used.

References: See the Notice of Availability on the inside front cover of this report to obtain a copy of:

1. NUREG-1556 Vol. 18, "Program-Specific Guidance About Service Provider Licenses," dated November 2000;

2. Regulatory Guide 8.25, Revision 1, "Air Sampling in the Workplace," dated June 1992; and

3. NUREG-1400, "Air Sampling in the Workplace," dated September 1993.

Additional References:

4. The Health Physics & Radiological Health Handbook, Third Edition, Edited by Bernard Shleien, dated 1998;

5. ANSI N323A-1997, "Radiation Protection Instrumentation Test and Calibration." Copies may be obtained from the American National Standards Institute (ANSI), 1430 Broadway, New York, NY 10018 or ordered electronically at the following address: www.ansi.org; and

6. "Air Sampling Instruments," American Conference of Governmental Industrial Hygienists, 9th Edition, dated 2001.

APPENDIX K

Public Dose

Public Dose

This Appendix describes different methods for determining radiation doses to members of the public.

Licensees must ensure that:

- The radiation doses received by individual members of the public do not exceed 1 millisievert (mSv) (100 millirem (mrem)) in one calendar year resulting from the licensee's possession and/or use of licensed materials (10 CFR 20.1301),
- Air emissions of radioactive material to the environment will not result in exposures to individual members of the public in excess of 0.1 mSv (10 mrem) (TEDE) in one year from those emissions (10 CFR 20.1101), and
- The radiation dose in unrestricted areas does not exceed 0.02 mSv (2 mrem) in any one hour (10 CFR 20.1301).

> Members of the public include persons who live, work, or may be near locations where byproduct material is used or stored and employees whose assigned duties do not include the use of byproduct material but may work in the vicinity where such materials are used or stored.

Doses to Members of the Public

INCLUDES doses from:	DOES NOT INCLUDE doses from:
• Radiation and/or radioactive material released by a licensee,	• Sanitary sewerage discharges from licensees,
• Sources of radiation which may or may not be licensed by NRC, under the control of a licensee, and	• Natural background radiation,
	• Medical administration of radioactive material, or
• Air effluents from sources of licensed radioactive materials.	• Voluntary participation in medical research.

> Typical unrestricted areas may include offices, shops, areas outside buildings, property, and storage areas. The licensee does not control access to these areas for purposes of controlling exposure to radiation or radioactive materials. However, the licensee may control access to these areas for other reasons, such as security.

The licensee may show compliance with the annual dose and constraint limits for individual members of the public by:

- Demonstrating by measurement or calculation that the TEDE to the individual likely to receive the highest dose at the boundary of the unrestricted area does not exceed 1 mSv (100 mrem) from all exposure pathways, and does not exceed 0.1 mSv (10 mrem) from air emissions; and

- Demonstrating that the annual average concentration of radioactive material released in gaseous and liquid effluents at the boundary of the unrestricted area does not exceed the values specified in Table 2 of Appendix B to 10 CFR Part 20 (20% of the values for gaseous effluents); and if an individual were continuously present in an unrestricted area, the dose from external sources would not exceed 0.02 mSv (2 mrem) in an hour and 0.5 mSv (0.05 rem) in a year.

In order to perform a dose assessment, the licensee should identify all potential sources of external and internal radiation exposure to members of the public and all locations of use, transport, and storage of radioactive material at its facility. The licensee must then take radiation measurements or perform calculations to demonstrate compliance.

Measurements

The licensee may use measurements to demonstrate that the TEDE to the individual likely to receive the highest dose at the boundary of the unrestricted area does not exceed 1 mSv (100 mrem) and does not exceed 0.1 mSv (10 mrem) from air emissions. These measurements may include:

- Dose rate surveys for radiation exposures from external radiation sources, and

- Measurements of radionuclides in air and water effluents.

The method used to measure dose will depend upon the nature of the radiation source. If the source of radiation is constant, it may be adequate to measure the dose rate and integrate it over time. If the source of radiation differs or changes over time, it may be necessary to perform continuous measurements.

Radioactivity releases may be determined by effluent monitoring or by effluent sampling and analysis. At radiopharmacies, airborne effluents are discharged when potentially volatile materials are used, such as during iodine capsule preparation, but the discharge itself is usually not continuous since volatile materials are used periodically rather than continuously. Liquid effluents may be discharged continuously or may be stored and subsequently discharged on a batch basis. For each type of source and for each route of potential exposure, consider the location of measurement points, whether continuous or periodic monitoring is required, the frequency of sampling and measurement, and any additional information. For discharges of airborne radionuclides, for example, it may be necessary to obtain information on the efficiency of filters and the air flow rate of the discharge system, as well as meteorological data and the distance to the nearest individual member of the public. Regulatory Guide 4.20, "Constraints on Release of Airborne Radioactive Materials to the Environment for Licensees Other Than Power Reactors," provides guidance on methods acceptable to NRC for compliance with the constraint on air emissions to the environment.

Calculation Method

Using a calculation method, the licensee must determine the highest dose an individual is likely to receive at the boundary of the unrestricted area. The licensee must take into account the individual's exposure from external sources and the concentration of radionuclides in gaseous and liquid releases. In practice, the licensee may wish to make conservative assumptions to simplify the dose calculation (See Figure K.1).

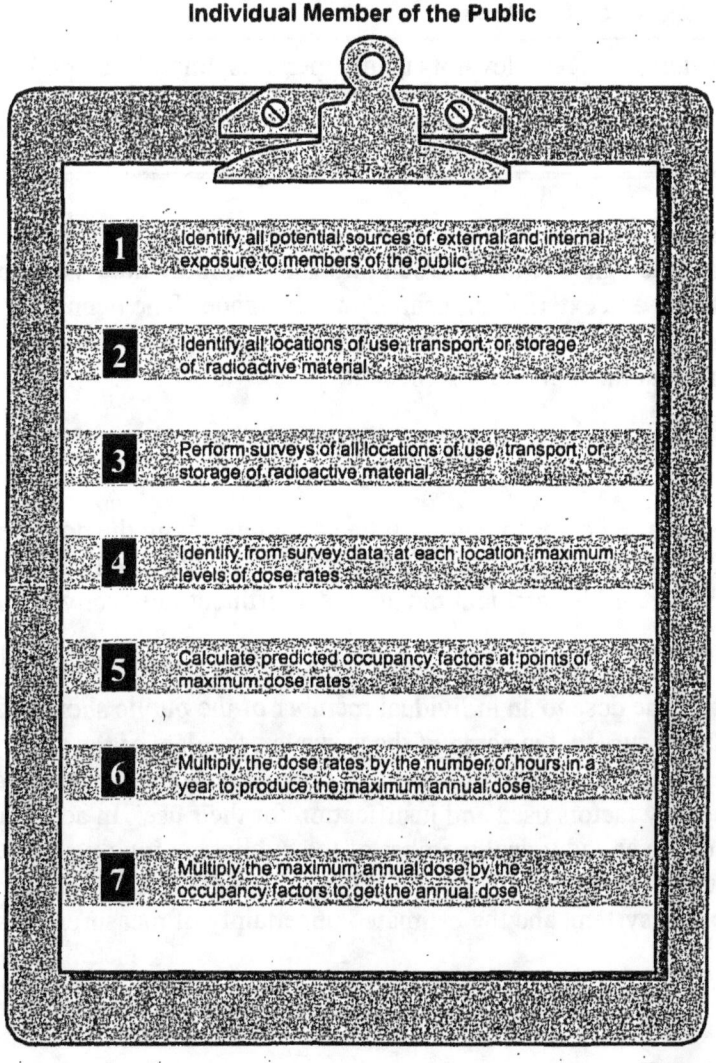

Calculating the Annual Dose to an Individual Member of the Public

1. Identify all potential sources of external and internal exposure to members of the public

2. Identify all locations of use, transport, or storage of radioactive material

3. Perform surveys of all locations of use, transport, or storage of radioactive material

4. Identify from survey data, at each location, maximum levels of dose rates

5. Calculate predicted occupancy factors at points of maximum dose rates

6. Multiply the dose rates by the number of hours in a year to produce the maximum annual dose

7. Multiply the maximum annual dose by the occupancy factors to get the annual dose

RPh18a.ppt
082898

Figure K.1 Calculating Public Dose. *Steps to calculate the annual dose to an individual member of the public.*

The public dose limit applies to the individual who is likely to receive the highest dose from licensed operations; therefore, the dose calculations must consider the location with the potential for the highest internal and external exposures. This calculation should assume that the individual was continuously present 24 hours a day, 365 days a year, or an occupancy factor of 1 (see Table K.1). If the result of the calculation using an occupancy factor of 1 demonstrates that the public dose and constraint limits are not exceeded, then there is no need for further evaluation.

Table K.1   Standard Occupancy Factors

Occupancy Factor	Description
1	Work areas such as offices, laboratories, shops, and occupied space in nearby buildings or outdoor areas
1/4	Corridors, lounges, elevators using operators, unattended parking lots
1/16	Waiting rooms, restrooms, stairways, unattended elevators, janitor's closets, outside areas used only for pedestrians or vehicular traffic

If the calculation demonstrates that either the public dose or constraint limit is exceeded with an occupancy factor of 1, then more realistic assumptions of the individual's occupancy at the points of highest internal and external exposures must be made. The licensee may use the occupancy factors in Table K.1 or may calculate a specific occupancy factor by determining the likely fraction of time that the individual is present.

Records

The licensee must maintain records to demonstrate compliance with the dose limit for individual members of the public, until the Commission terminates the license. In general, survey and monitoring records of ambient radiation and effluent radioactivity should be adequate.

Records demonstrating the dose to an individual member of the public should identify the instrument(s) used in the survey, the name of the surveyor, the date of the survey, the location of the survey(s), including a description or drawing of the area surveyed, survey results, and, if applicable, the occupancy factors used and justification for their use. In addition, records demonstrating the dose to an individual member of the public that involve effluent sampling analysis should include information on concentrations of specific radionuclides, the minimum detectable activity of the system, and the estimated uncertainty of measurements.

APPENDIX L

Model Leak Test Program

Model Leak Test Program

Training

Before allowing an individual to perform leak testing, the licensee must ensure that he or she has sufficient classroom and on-the-job training to show competency in performing leak tests independently.

Classroom training may be in the form of lecture, videotape, or self-study and will cover the following subject areas:

- Principles and practices of radiation protection;

- Radioactivity measurements, monitoring techniques, and using instruments;

- Mathematics and calculations used for measuring radioactivity; and

- Biological effects of radiation.

Appropriate on-the-job-training consists of:

- Observing authorized personnel collecting and analyzing leak-test samples, and

- Collecting and analyzing leak-test samples under the supervision and in the physical presence of an individual authorized to perform leak tests.

Facilities and Equipment

- To ensure achieving the required sensitivity of measurements, leak tests will be analyzed in a low-background area.

- Use a calibrated and operable survey instrument to check leak-test samples for gross contamination before they are analyzed.

- Analyze the leak-test sample using an instrument that is appropriate for the type of radiation to be measured (e.g., NaI (Tl) well-counter system for gamma-emitters, liquid scintillation for beta-emitters, gas-flow proportional counters for alpha-emitters).

- If the sensitivity of the counting system is unknown, the minimum detectable activity (MDA) needs to be determined. The MDA may be determined using the following formula:

$$MDA = \frac{2.71 + 4.65 \sqrt{(B_R \times t)}}{t \times E} = \text{Minimum Detectable Activity}$$

where: MDA = minimum detectable activity in disintegrations per minute (dpm)
       bkg   = background count rate in counts per minute (cpm)
       t   = background counting time in minutes
       E   = detector efficiency in counts per disintegration

For example:

where: bkg   =   200 counts per minute (cpm)
         E   =   0.1 counts per disintegration (10% efficient)
         t   =   2 minutes

$$MDA = \frac{2.71 + 4.65 \sqrt{(200 \text{ cpm} \times 2 \text{ minutes})}}{2 \times 0.1} = \frac{2.71 + 4.65\sqrt{(400)}}{0.2}$$

$$= \frac{2.71 + 4.65 (20)}{0.2} = \frac{2.71 + 93}{0.2} = \frac{95.71}{0.2}$$

$$= \frac{478.55 \text{ disintegrations}}{\text{minute}}$$

$$\text{becquerels (Bq)} = \frac{1 \text{ disintegration}}{\text{second}}$$

$$Bq = \frac{478.55 \text{ disintegration}}{\text{minutes}} \times \frac{\text{minute}}{60 \text{ seconds}} = 7.976 \text{ Bq}$$

Frequency for Conducting Leak Tests of Sealed Sources

Leak tests will be conducted at the frequency specified in the respective SSDR certificate.

Procedure for Performing Leak Testing and Analysis

- For each source to be tested, list identifying information such as sealed source serial number, radionuclide, activity.

- If available, use a survey meter to monitor exposure.

- Prepare a separate wipe sample (e.g., cotton swab or filter paper) for each source.

- Number each wipe to correlate with identifying information for each source.

- Wipe the most accessible area where contamination would accumulate if the sealed source were leaking.

- Select an instrument that is sensitive enough to detect 185 Bq (0.005 microcuries) of the radionuclide.

- Using the selected instrument, count and record background count rate.

- Check the instrument's counting efficiency using a standard source of the same radionuclide as the source being tested or one with similar energy characteristics. Accuracy of standards should be within ± 5% of the stated value and traceable to primary radiation standards such as those maintained by the NIST.

- Calculate efficiency.

For example: [(cpm from std) - (cpm from bkg)] = efficiency in cpm/Bq
               activity of std in Bq

    where: cpm   =   counts per minute
              std   =   standard
              bkg   =   background
              Bq   =   becquerel

- Count each wipe sample; determine net count rate.

- For each sample, calculate and record estimated activity in Bq (or mCi).

For example: [(cpm from wipe sample) - (cpm from bkg)] = Bq on wipe sample
               efficiency in cpm/Bq

- Sign and date the list of sources, data, and calculations. Retain records for 3 years
  (10 CFR 20.2103(a)). If the wipe test activity is 185 Bq (0.005 microcurie) or greater, notify
  the RSO, so that the source can be withdrawn from use and disposed of properly. Also notify
  NRC.

Reference: See NUREG-1556, Vol. 18, "Program-Specific Guidance About Service Provider
Licenses," dated November 2000.

APPENDIX M

Summary of DOT Requirements for Transportation of Type A or Type B Quantities of Licensed Material

Summary of DOT Requirements for Transportation of Type A or Type B Quantities of Licensed Material

Licensed material must be transported in accordance with DOT regulations. The major areas in the DOT regulations that are most relevant for transportation of Type A or Type B quantities of licensed material are:

- Table of Hazardous Materials and Special Provisions - 49 CFR 172.101: Purpose and use of hazardous materials table;

- Shipping Papers - 49 CFR 172.200-204: Applicability, general entries, description of hazardous material on shipping papers, additional description requirements, shipper's certification;

- Package Marking - 49 CFR 172.300, 49 CFR 172.301, 49 CFR 172.303, 49 CFR 172.304, 49 CFR 172.310, 49 CFR 172.324: Applicability, general marking requirements for nonbulk packagings, prohibited marking, marking requirements, radioactive material, hazardous substances in nonbulk packaging;

- Package Labeling - 49 CFR 172.400, 49 CFR 172.401, 49 CFR 172.403, 49 CFR 172.406, 49 CFR 172.407, 49 CFR 172.436, 49 CFR 172.438, 49 CFR 172.440: General labeling requirements, prohibited labeling, Class 7 (radioactive) material, placement of labels, label specifications, radioactive white-I label, radioactive yellow-II label, radioactive yellow-III label;

- Placarding of Vehicles - 49 CFR 172.500, 49 CFR 172.502, 49 CFR 172.504, 49 CFR 172.506, 49 CFR 172.516, 49 CFR 172.519, 49 CFR 172.556: Applicability of placarding requirements, prohibited and permissive placarding, general placarding requirements, providing and affixing placards: highway, visibility and display of placards, general specifications for placards, RADIOACTIVE placard;

- Emergency Response Information - 49 CFR 172.600, 49 CFR 172.602, 49 CFR 172.604: Applicability and general requirements, emergency response information, emergency response telephone number;

- Training - 49 CFR 172.702, 49 CFR 172.704: Applicability and responsibility for training and testing requirements;

- Shippers – General Requirements for Shipments and Packaging - 49 CFR 173.403, 49 CFR 173.410, 49 CFR 173.411, 49 CFR 173.412, 49 CFR 173.413, 49 CFR 173.415, 49 CFR 173.416, 49 CFR 173.433, 49 CFR 173.435, 49 CFR 173.441, 49 CFR 173.471, 49 CFR 173.475, 49 CFR 173.476: Definitions, general design requirements, industrial packages, additional design requirements for Type A packages, requirements for Type B packages, authorized Type A packages, authorized Type B packages, requirements for determining A1 and A2 values for radionuclides and for the listing of radionuclides on shipping papers and labels, table of A1 and A2 values for radionuclides, radiation level limitations, requirements for NRC-approved packages, quality control requirements prior to each shipment of Class 7 (radioactive) materials, approval of special form Class 7 (radioactive) materials; and

- Carriage by Public Highway - 49 CFR 177.816, 49 CFR 177.817, 49 CFR 177.834(a), 49 CFR 177.842: Driver training, shipping papers, general requirements (packages secured in a vehicle), Class 7 (radioactive) material.

For additional transportation information, licensees may consult DOT's "A Review of the Department of Transportation Regulations for Transportation of Radioactive Materials" or go to the DOT web page site at http://hazmat.dot.gov.

APPENDIX N

Model Personnel Training Program

Model Personnel Training Program

Training Program

1.　General instructions

　　1.1　Training will be provided:

- Before an employee assumes duties with or in the immediate vicinity of radioactive materials,

- At least annually, as refresher training for all employees, and

- Whenever a significant change occurs in duties, regulations, or the terms of an NRC license.

　　1.2　Subjects covered for individuals working with, or in the vicinity of, radioactive materials or radiation:

- Safe radiation practices associated with the job (examples of topics that may be covered are found in Section 3 of this Appendix),

- Site-specific radiation safety practices, and

- Applicable NRC regulations.

　　1.3　Subjects covered for ancillary personnel:

- Significance of the radiation symbol and its use on signs and labels,

- Location of unrestricted areas, and

- Whether the individual is authorized access to the restricted areas of the facility.

　　1.4　Type of instruction:

- Instruction in the licensee's site-specific Radiation Safety Program and NRC regulatory requirements may be in the form of lecture, demonstrations, videotape, or self-study, and should emphasize practical subjects important to the safe use of licensed material, and

- Individuals receiving instructions should be provided an opportunity to ask questions.

2.　Instruction for individuals likely to receive an occupational dose in excess of 1 mSv (100 mrem)

　　2.1　Instruction will be provided:

- Before an employee assumes duties with or in the immediate vicinity of radioactive materials,

- At least annually, as refresher training, and

- Whenever a significant change occurs in duties, regulations, or terms of NRC license.

2.2 Licensee must provide instruction in subjects covered in 10 CFR 19.12.

2.3 Records of initial and refresher training should be maintained and should include:

- Name of the individual who provided the instruction,

- Names of the individuals who received the instruction,

- Date of instruction, and

- List of the topics covered.

3. Suggested radiation safety training topics for individuals working with, or in the vicinity of, byproduct material (this section is intended as a guide to topics covered in a typical radiation safety training program; topics selected should be commensurate with the individuals' duties).

3.1 Basic radiation safety information:

- Basic radiation biology (e.g., interaction of ionizing radiation with cells and tissues),

- Radiation safety

  — radiation vs. contamination,

  — internal vs. external exposure,

  — biological effects of radiation,

  — ALARA concept, and

  — use of time, distance, and shielding to minimize exposure,

- Risk estimates, including comparison with other health risks (10 CFR 19.12),

- Regulatory requirements

  — RSO,

  — material control and accountability,

  — dose to individual members of the public,

  — personnel dosimetry,

  — occupational dose limits and their significance,

  — dose limits to the embryo/fetus, including instruction on declaration of pregnancy,

  — workers' right to be informed of occupational radiation exposure,

  — Radiation Safety Program audits,

— ordering and receipt of packages,

— transfer,

— waste disposal,

— recordkeeping,

— surveys,

— postings,

— labeling of containers,

— handling and reporting of incidents or events,

— licensing and inspection by NRC,

— need for complete and accurate information,

— employee protection, and

— deliberate misconduct.

3.2 General topics for safe use of radionuclides:

• Wear a laboratory coat or other protective clothing at all times when working with radioactive materials.

• Use syringe shields and vial shields when preparing and handling radioactive drugs.

• Measure all radiopharmaceuticals prior to transfer.

• Measure the molybdenum-99 content of each generator elution and do not transfer those radiopharmaceuticals for human medical use that will contain more than 0.15 microcuries of molybdenum-99 per mCi of technetium-99m at the time of administration.

• Wear disposable gloves at all times when handling radioactive materials and change gloves frequently to minimize the spread of contamination.

• Before leaving the hot lab, monitor hands, shoes, and clothing for contamination in a low-background area, allowing sufficient time for instrument response.

• Do not eat, drink, smoke, or apply cosmetics in any area where licensed material is stored or used.

• Do not store food, drink, or personal effects in areas where licensed material is stored or used (see Figure Q.1). Personal items brought into the restricted area (radios, compact discs, notepads, books, etc.) should be surveyed for contamination before removal from the area.

• Food and beverages used in the preparation of radiopharmaceuticals should be clearly labeled "Not for personal consumption" if stored with radioactive materials.

• Wear personnel monitoring devices, if required, at all times while in areas where licensed materials are used or stored.

- Dispose of radioactive waste only in designated, labeled, and properly shielded receptacles.
- Never pipette by mouth.
- Store radioactive solutions in clearly labeled containers.
- Secure all licensed material when it is not under the constant surveillance and immediate control of the user(s).

3.3 Instruction on radiopharmacy-specific program elements:

- Applicable regulations and license conditions,
- Areas where radioactive material is used or stored,
- Potential hazards associated with radioactive material in each area where the individuals will work,
- Special procedures for handling volatile materials,
- Proper use of radiation shielding,
- Proper use of survey and analytical instruments,
- Appropriate response to spills, emergencies, or other unsafe conditions,
- Emergency procedures,
- Previous incidents, events, and/or accidents,
- Survey program,
- Effluent monitoring and control,
- Customer-returned waste pickup, receipt, and handling,
- Waste management and minimization,
- Personnel monitoring,
- Procedures for receiving packages containing radioactive materials,
- Procedures for opening packages,
- Sealed sources and leak tests, and
- Other topics, as applicable.

APPENDIX O

Model Dose Calibrator Testing Program

Model Dose Calibrator Testing Program

Model Procedures for Testing Dose Calibrators Used to Measure Photon-Emitting Radionuclides

This model procedure can be used by applicants and licensees for checking and testing dose calibrators.

Model Procedure

1.   Test for the following at the indicated frequency (consider repair, replacement, or arithmetic correction if the dose calibrator falls outside the suggested tolerances):

   1.1   Constancy, at least once each day prior to assay of patient dosages (a safe margin is considered to be below ±10%),

   1.2   Linearity at installation and at least quarterly thereafter (a safe margin is considered to be below ±10%),

   1.3   Geometry dependence at installation (a safe margin is considered to be below ±10%), and

   1.4   Accuracy, at installation and at least annually thereafter (a safe margin is considered to be below ±10%).

2.   After repair, adjustment, or relocation of the dose calibrator, such that proper function of the ionization chamber or electronics would likely be in doubt, repeat the above tests as appropriate.

3.   Constancy means reproducibility in measuring a constant source over a long period of time. Assay at least one relatively long-lived source such as cesium-137, cobalt-60, cobalt-57, or radium-226, using a reproducible geometry each day before using the calibrator; consider using two or more sources with different photon energies and activities.

   Use the following procedure:

   3.1   Assay each reference source using the appropriate dose calibrator setting (e.g., use the cesium-137 setting to assay cesium-137).

   3.2   Measure background at the same setting, and subtract or confirm the proper operation of the automatic background circuit if it is used.

   3.3   For each source used, either plot or log (i.e., record in the dose calibrator log book) the background level for each setting checked and the net activity of each constancy source.

3.4 Using one of the sources, repeat the above procedure for all commonly used radionuclide settings. Plot or log the results.

3.5 Establish an action level or tolerance for each recorded measurement at which the individual performing the test will automatically notify the ANP or the RSO of a suspected malfunction of the calibrator. These action levels should be written in the log book or posted on the calibrator. The dose calibrator should be repaired or replaced if the error exceeds 10%.

4. The linearity of a dose calibrator should be ascertained over the range of its use between the maximum activity in a vial and 30 microcuries. Note that with radionuclides with short half-lives such as PET radionuclides, there may be difficulties measuring a low activity such as 30 microcuries. Therefore, the lowest activity that is measurable, which must be below the lowest dose distributed, is acceptable. Linearity means that the calibrator is able to indicate the correct activity over the range of use of that calibrator. This example uses a vial of technetium-99m that has the anticipated maximum activity to be assayed (e.g., the first elution from a new generator) and assumes the predetermined safety margin is ±5%.

4.1 Time Decay Method

4.1.1 Inspect the instrument to ascertain that the measurement chamber liner is in place and that instrument zero is properly set (see manufacturer's instructions).

4.1.2 Assay the technetium-99m vial in the dose calibrator and subtract background to obtain net activity in millicuries.

4.1.3 Repeat step in Section 4.1.2 at time intervals of 6, 24, 30, and 48 hours after the initial assay.

Note: Time intervals used for other radionuclides may vary depending on the radionuclide's half-life.

4.1.4 Using the 30-hour activity measurement as a starting point, calculate the predicted activities at 0, 6, 24, and 48 hours using the following table:

Assay Time[1] (hours)	Correction Factor
0	31.6
6	15.8
24	2.00
30	1.00
48	0.126

---

[1] Assay times should be measured in whole hours and correction factors should be used to three significant figures as indicated. The half-life of $T_{1/2} = 6.02$ hours has been used in calculating these correction factors.
Example: If the net activity measured at 30 hours was 15.6 mCi, the calculated activities for 6 and 48 hours would be 15.6 mCi x 15.9 = 248 mCi and 15.6 mCi x 0.126 = 1.97 mCi, respectively.

4.1.5   Plot both the measured net activity and the calculated activity versus time.

4.1.6   On the graph, the measured net activity plotted should be within ±5% of the calculated activity if the instrument is linear and functioning properly. If variations greater than 5% are noted, adjust the instrument, have it repaired, or use arithmetic correction factors to correct the readings obtained in daily operations.

4.1.7   If instrument linearity cannot be corrected, for routine assays it will be necessary to use either an aliquot of the eluate that can be accurately measured or the graph constructed in Section 4.1.5 to relate measured activities to calculated activities.

4.2   Shield Method: If a set of "sleeves" of various thicknesses are used to test for linearity, it will first be necessary to calibrate them.

4.2.1   Begin the linearity test by assaying the technetium-99m syringe or vial in the dose calibrator, and subtract background to obtain the net activity in millicuries. Record the date, time to the nearest minute, and net activity. This first assay should be done in the morning at a regular time. After making the first assay, the sleeves can be calibrated as follows. (Steps in Sections 4.2.2 through 4.2.4 must be completed within 6 minutes.)

4.2.2   Put the base and sleeve 1 in the dose calibrator with the vial. Record the sleeve number and indicated activity.

4.2.3   Remove sleeve 1 and put in sleeve 2. Record the sleeve number and indicated activity.

4.2.4   Continue for all sleeves.

4.2.5   Complete the following decay-method linearity test steps:

4.2.5.1   Repeat the assay at about noon, and again at about 4:00 p.m. Continue on subsequent days until the assayed activity is less than 30 microcuries. For dose calibrators on which the range is selected with a switch, select the range normally used for the measurement.

4.2.5.2   Convert the time and date information recorded to hours elapsed since the first assay.

4.2.5.3   On a sheet of semilog graph paper, label the logarithmic vertical axis in millicuries and label the linear horizontal axis in hours elapsed. At the top of the graph, note the date and the manufacturer, model number, and serial number of the dose calibrator. Plot the data.

4.2.5.4   Draw a "best fit" straight line through the data points. For the point farthest from the line, calculate its deviation from the value on the line.
(A-observed - A-line)/(A-line) = deviation

4.2.5.5   If the worst deviation is more than ±0.05, the dose calibrator should be repaired or adjusted. If this cannot be done, it will be necessary to make a correction table or graph that will allow conversion from activity indicated by the dose calibrator to "true activity."

4.2.6   From the graph made in Section 4.2.5.3, find the decay time associated with the activity indicated with sleeve 1 in place. This is the "equivalent decay time" for sleeve 1. Record that time with the data recorded in Section 4.2.2.

4.2.7   Find the decay time associated with the activity indicated with sleeve 2 in place. This is the "equivalent decay time" for sleeve 2. Record that time with the data recorded in Section 4.2.3.

4.2.8   Continue for all sleeves.

4.2.9   The table of sleeve numbers and equivalent decay times constitutes the calibration of the sleeve set.

The sleeve set may now be used to test dose calibrators for linearity

4.2.10   Assay the technetium-99m syringe or vial in the dose calibrator, and subtract background to obtain the net activity in millicuries. Record the net activity.

4.2.11   Steps in Section 4.2.12 through 4.2.14 below must be completed within 6 minutes.

4.2.12   Put the base and sleeve 1 in the dose calibrator with the vial. Record the sleeve number and indicated activity.

4.2.13   Remove sleeve 1 and put in sleeve 2. Record the sleeve number and indicated activity.

4.2.14   Continue for all sleeves.

4.2.15   On a sheet of semilog graph paper, label the logarithmic vertical axis in millicuries, and label the linear horizontal axis in hours elapsed. At the top of the graph, note the date and the model number and serial number of the dose calibrator.

4.2.16 Plot the data using the equivalent decay time associated with each sleeve.

4.2.17 Draw a "best fit" straight line through the data points. For the point farthest from the line, calculate its deviation from the value on the line. (A-observed - A-line)/(A-line) = deviation.

4.2.18 If the worst deviation is more than ±0.05, the dose calibrator should be repaired or adjusted. If this cannot be done, it will be necessary to make a correction table or graph that will allow conversion from activity indicated by the dose calibrator to "true activity."

5. Geometry independence means that the indicated activity does not change with volume or configuration. The test for geometry independence should be conducted using syringes and vials that are representative of the entire range of size, shape, and constructions normally used for injections and a vial similar in size, shape, and construction to the radiopharmaceutical kit vials normally used. The following example assumes that injections are done with 3-cc plastic syringes, that radiopharmaceutical kits are made in 30-cc glass vials, and that the predetermined safety margin is ±5%.

5.1 In a small beaker or vial, mix 2 cc of a solution of technetium-99m with an activity concentration between 1 and 10 mCi/ml. Set out a second small beaker or vial with nonradioactive saline. Tap water may be used.

5.2 Draw 0.5 cc of the technetium-99m solution into the syringe and assay it. Record the volume and millicuries.

5.3 Remove the syringe from the calibrator, draw an additional 0.5 cc of nonradioactive saline or tap water, and assay again. Record the volume and millicuries indicated.

5.4 Repeat the process until a volume of 2.0 cc has been assayed. The entire process must be completed within 10 minutes.

5.5 Select as a standard the volume closest to that normally used for injections. For all other volumes, divide the standard millicuries by the millicuries indicated for each volume. The quotient is a volume correction factor. Alternatively, graph the data and draw horizontal error lines above and below the chosen "standard volume."

5.6 If any correction factors are greater than 1.05 or less than 0.95, or if any data points lie outside the error lines, it will be necessary to make a correction table or graph that will allow a conversion from "indicated activity" to "true activity." If this is necessary, be sure to label the table or graph "syringe geometry dependence," and note the date of the test and the model and serial number of the calibrator.

5.7 To test the geometry dependence for a 30-cc glass vial, draw 1.0 cc of the technetium-99m solution into a syringe and then inject it into the vial. Assay the vial. Record the volume and millicuries indicated.

5.8 Remove the vial from the calibrator and, using a clean syringe, inject 2.0 cc of nonradioactive saline or tap water, and assay again. Record the volume and millicuries indicated.

5.9 Repeat the process until a volume of 19.0 cc has been assayed. The entire process must be completed within 10 minutes.

5.10 Select as a standard the volume closest to that normally used for mixing radiopharmaceutical kits. For all other volumes, divide the standard millicuries by the millicuries indicated for each volume. The quotient is a volume correction factor. Alternatively, the data may be graphed, with horizontal 5% error lines drawn above and below the chosen "standard volume."

5.11 If any correction factors are greater than 1.05, or less than 0.95, or if any data points lie outside the 5% error lines, it will be necessary to make a correction table or graph that will allow conversion from "indicated activity" to "true activity." If this is necessary, be sure to label the table or graph "vial geometry dependence," and note the date of the test and the model number and serial number of the calibrator.

6. Accuracy means that, for a given calibrated reference source, the indicated millicurie value is equal to the millicurie value determined by the NIST or by the supplier who has compared that source to a source that was calibrated by NIST. Certified sources are available from NIST and from many radionuclide suppliers. At least two sources with different principal photon energies (such as cobalt-57, cobalt-60, cesium-137) should be used. One source should have a principal photon energy between 100 keV and 500 keV. If a radium-226 source is used, it should be at least 10 microcuries; other sources should be at least 50 microcuries. Consider using at least one reference source with an activity that is within the range of activities normally assayed.

6.1 Assay a calibrated reference source at the appropriate setting (e.g., use the cobalt-57 setting to assay cobalt-57) and then remove the source and measure background. Subtract background from the indicated activity to obtain the net activity. Record this measurement. Repeat for a total of three determinations.

6.2 Average the three determinations. The average value should be within the predetermined safety margin, which in this example is 5% of the certified activity of the reference source, mathematically corrected for decay.

6.3 Repeat the procedure for other calibrated reference sources.

6.4   If the average value does not agree, within 5%, with the certified value of the reference source, the dose calibrator may need to be repaired or adjusted. The dose calibrator should be repaired or replaced if the error exceeds 10%.

6.5   At the same time the accuracy test is performed, assay the source that will be used for the daily constancy test (it need not be a certified reference source) on all commonly used radionuclide settings. Record the settings and indicated millicurie values with the accuracy data.

6.6   Put a sticker on the dose calibrator, noting when the next accuracy test is due.

7.   The individual performing the tests will sign or initial the records of all geometry, linearity, and accuracy tests.

APPENDIX P

Material Receipt and Accountability

Material Receipt and Accountability

Sample Model Procedure for Ordering and Receiving Radioactive Material

- The RSO should approve or place all orders for radioactive material and should ensure that the requested material, quantities, manufacturer, and model are authorized by the license and that the possession limits are not exceeded.

- Carriers should be instructed to deliver radioactive packages directly to the designated receiving area.

Sample Instructions to Personnel Involved in Material Receipt

Shipping and Receiving Personnel

During normal working hours, within 3 hours of receipt of any package of licensed material, each package must be visually inspected for any signs of shipping damage, such as crushed or punctured containers or signs of dampness. Any suspected damage must be reported to the RSO immediately. Do not touch any package suspected of leaking. Request the person delivering the package, if still on site, to remain until monitored by the RSO.

Outside of normal working hours (e.g., nights, weekends, and holidays), deliveries may be made to a designated, secured storage area. These packages must be checked for contamination and external radiation levels within 3 hours after personnel arrive at the facility. They should not be allowed to remain in the designated storage area any longer than necessary, as they may be a source of exposure for personnel.

Sample Model Procedure for Safely Opening Packages Containing Licensed Materials

For packages received under the specific license, authorized individuals should implement procedures for opening each package, as follows:

- Wear gloves to prevent hand contamination.

- Visually inspect the package for any sign of damage (e.g., crushed, punctured). If damage is noted, stop and notify the RSO.

- Check DOT White I, Yellow II, or Yellow III label or packing slip for activity of contents, to ensure that the shipment does not exceed license possession limits.

- Monitor the external surfaces of a labeled package according to specifications in Table 8.1.

- Open the outer package (following supplier's directions if provided) and remove packing slip. Open inner package to verify contents, comparing requisition, packing slip, and label on the container. Check integrity of the final source container (e.g., inspecting for breakage of seals or vials, loss of liquid, discoloration of packaging material, high count rate on smear). Again check that the shipment does not exceed license possession limits. If anything other than the expected observation is identified, stop and notify the RSO.

- Survey the packing material and packages for contamination before discarding. If contamination is found, treat as radioactive waste. If no contamination is found, obliterate the radiation labels prior to discarding in the regular trash.

- Maintain records of receipt, package survey, and wipe test results.

- Notify the final carrier and the NRC Operations Center when removable radioactive surface contamination exceeds the limits of 22 disintegrations per minute per square centimeter (dpm/cm^2) averaged over 300 cm^2, or external radiation levels exceed 2.0 mSv/hr (200 mrem/hr) at the surface.

APPENDIX Q

General Topics for Safe Use of Radionuclides and Model Emergency Procedures

General Topics for Safe Use of Radionuclides and Model Emergency Procedures

General Topics for Safe Use of Radionuclides

Each licensee using radioactive material should establish general rules for the safe use of the material so that workers know what is required. Typical instructions should include:

- Wear a laboratory coat or other protective clothing at all times when working with radioactive materials.

- Use syringe shields and vial shields when preparing and handling radioactive drugs.

- Measure all radiopharmaceuticals prior to transfer.

- Measure the molybdenum-99 content of the first generator elution and do not transfer those radiopharmaceuticals for human medical use that will contain more than 0.15 microcuries of molybdenum-99 per millicurie of technetium-99m at the time of administration.

- Wear disposable gloves at all times when handling radioactive materials and change gloves frequently to minimize the spread of contamination.

- Before leaving the hot lab, monitor hands, shoes, and clothing for contamination in a low-background area, allowing sufficient time for instrument response.

- Do not eat, drink, smoke, or apply cosmetics in any area where licensed material is stored or used.

- Do not store food, drink, or personal effects in areas where licensed material is stored or used (see Figure Q.1). Personal items brought into the restricted area (radios, compact discs, notepads, books, etc.) should be surveyed for contamination before removal from the area.

- Food and beverages used in the preparation of radiopharmaceuticals should be clearly labeled "Not for personal consumption" if stored with radioactive materials.

- Wear personnel monitoring devices, if required, at all times while in areas where licensed materials are used or stored.

- Dispose of radioactive waste only in designated, labeled, and properly shielded receptacles.

- Never pipette by mouth.

- Store radioactive solutions in clearly labeled containers.

- Secure all licensed material when it is not under the constant surveillance and immediate control of the user(s).

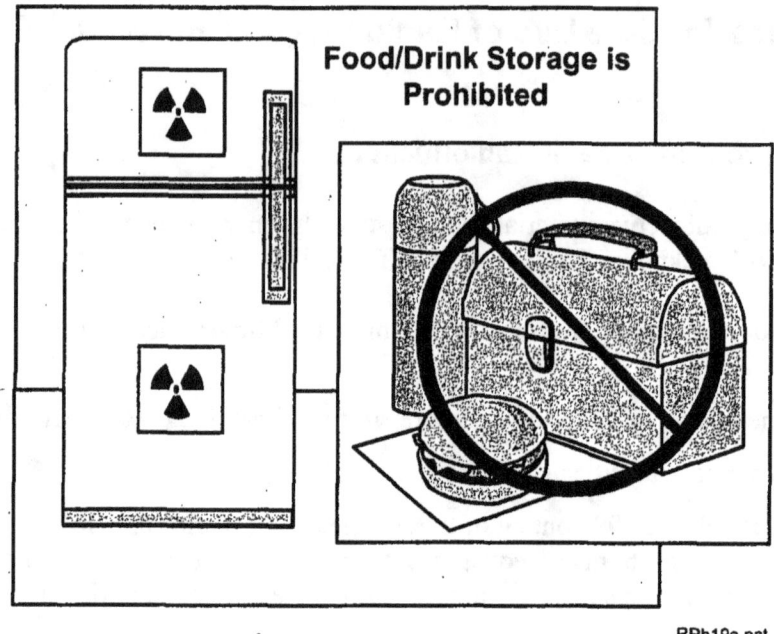

RPh19a.ppt
082898

Figure Q.1 Storage of Food and Drink. *Food or drink for personal consumption should not be stored in refrigerators with radionuclides.*

Model Procedures for Handling Millicurie Quantities of Radioiodine

Because of the potential for significant intakes due to volatility and accidental ingestion, and the potential for skin exposures (shallow-dose equivalent) from contamination, licensees should establish specific procedures for the containment and handling of millicurie quantities of radioiodine, most commonly iodine-131. The following guidance is the minimum that should be considered if the applicant intends to manipulate radioiodine.

- Manipulation of radioiodine (e.g., handling or compounding capsules, performing radioiodination, dispensing from bulk solution) should be conducted in an isolated area within the main hot lab of the pharmacy. This will aid in maintaining exposures ALARA and provide a means to isolate the area in the event of a spill.

- Radioiodine handling should only be performed inside a glovebox or fume hood. The ventilation for gloveboxes and fume hoods should be checked at least once every six months to ensure adequate airflow and confirm negative pressure with respect to the area around the glovebox or fume hood. Exhaust stacks for gloveboxes and fume hoods used for handling radioiodine should not be located near ventilation intakes to minimize the likelihood of recirculation to the pharmacy or to other tenants in a shared building.

- Gloveboxes and fume hoods must include appropriate filters (activated charcoal) to minimize effluents from radioiodine handling.

- Filters must be installed and used in accordance with the manufacturer's specifications (e.g., adequate air flow to ensure adequate residence time).

- Filters should be checked at installation and periodically, based on use, but not less than once per calendar quarter, to ensure continued efficiency.

- Air flow through fume hoods and gloveboxes should be confirmed before each use.

- Magna-helic sensors, if used, should be checked before each use of the glovebox or fume hood, to ensure minimum flow across the filter.

- Absorbent materials and dry chemical buffers, for use in the event of a spill, should be located near the area where millicurie quantities of radioiodine are handled.

- Additional protective clothing should be used when handling millicurie quantities of radioiodine. Personnel should be double-gloved and use shoulder-length sleeve guards. The gloves and glove seals on gloveboxes should be checked periodically and replaced when needed.

- All personnel handling greater than 500 millicuries of iodine-131 in a year should be considered for bioassay. This is the threshold below which intakes over 1% of the annual limit on intake (ALI) are not likely, and assumes no containment. When used in a properly operating fume hood, the threshold for consideration of the need for bioassay rises to 5 curies of iodine-131. If used in a properly operating glovebox, with properly sealed glove ports and well maintained gloves, the threshold rises to 50 curies of iodine-131 handled by one person per year. Pharmacies using gloveboxes that do not have sealed glove ports may not use the threshold indicated for that equipment, but may use the threshold for properly maintained fume hoods.

Model Procedures for Handling Events

Suggested Thresholds for Defining Minor Contamination Events, Minor Spills, and Major Spills

Licensees should establish clearly delineated thresholds for describing these types of events. Licensees should establish a graded response to emergencies, incorporating increasing formality of a response based on the potential risks posed by the events. No emergency procedure can anticipate every likely event; therefore, flexibility and judgment must be incorporated into such procedures. Most importantly, if licensee staff are not sure of the proper or expected response to any event, no matter how minor, they should be instructed to immediately cease further action, control access to the area, contact the RSO, and wait for instructions.

Although the following is only suggested guidance for establishing response thresholds, *significant* deviations in actual licensee emergency procedures should be clearly justified.

Minor Contamination Events

Minor contamination events are those events typically identified through routine surveys that involve removable contamination levels greater than the licensee's action limit, but less than ten times the licensee's action limit. These events can be easily decontaminated without the need for strict adherence to a step-by-step procedure. Such events require judgment on the part of the individuals responding to determine the scope and extent of the contamination and to assess their ability to respond effectively. In order to prevent the spread of contamination, coworkers should be notified if decontamination of the area will be delayed. The RSO should be notified promptly of such events, either before, or immediately after, cleanup of the area. Isolated minor contamination events may not require a formal root cause evaluation or extensive corrective action determinations; however, several events in the same location, involving the same individual, or during similar processes, may warrant such in-depth evaluations and determinations.

Minor Spills

Minor spills are those events typically identified at the time they occur (e.g., a dropped syringe or vial containing radioactive material) involving the release (spill) of radioactive material requiring a more formal adherence to a step-by-step procedure. Such events will usually involve millicurie quantities of material and have a potential for exposures to personnel or the public if not properly controlled and decontaminated. The upper limit for defining minor spills should not be more than five times the lowest ALI of the material involved in the spill. Such a limit would include the following quantities of radioactive material:

1. Up to 400 millicuries of technetium-99m,

2. Up to 150 microcuries of iodine-131,

3. Up to 250 millicuries of fluorine-18,

4. Up to 100 millicuries of thallium-201, and

5. Up to 10 millicuries of samarium-153.

Minor spills may warrant root cause evaluations and corrective action determinations, depending on the circumstances. The RSO should be notified immediately of such events so that decontamination procedures can be monitored. Minor spills involving quantities of radioactive material near the upper threshold may require more than one person to respond to assist in the cleanup, perform confirmation surveys, or monitor materials and personnel exiting the area.

Major Spills

Any spill involving a quantity of radioactive material in excess of the quantity defined for a minor spill is considered a major spill. Such spills have a greater potential for exposures to workers and the public, including the possibility of overexposure, if not properly contained. Individuals should never attempt to clean a major spill by themselves, or without the personal supervision and direction of the RSO. Major spills should generally be reported to NRC in accordance with the requirements of 10 CFR 30.50. Major spills may also require evaluations of intakes and skin doses, if personnel contamination is identified, as well as root cause evaluations and corrective action determinations. Qualified assistance should be sought immediately for those major spills that are beyond the licensee's capability to address.

General Safety Procedures to Handle Spills

- The name and telephone number of the RSO or an alternate person(s) should be posted conspicuously in areas of use, so that they are readily available to workers in case of emergencies. Licensees should have emergency equipment readily available for handling spills. Spill response materials should include the following:

  — disposable gloves,

  — housekeeping gloves,

  — disposable lab coats,

  — disposable shoe covers,

  — roll of absorbent paper with plastic backing,

  — masking tape,

  — plastic trash bags with twist ties,

  — "Radioactive Material" labeling tape,

  — marking pen,

  — pre-strung "Radioactive Material" labeling tags,

  — box of wipes,

  — instructions for "Emergency Procedures,"

  — clipboard with a copy of the Radioactive Spill Report Form for the facility, and

  — pen or pencil.

Minor Contaminations and Spills of Liquids and Solids

- Instructions to Workers

  — These instructions apply to minor contamination events (less than 10 times the licensee's action limit) and minor spills of radioactive material. The response to each is similar; however, the response to minor contamination events need not be as formal as the response to spills involving millicurie quantities of radioactive material.

  — Notify persons in the area that a spill has occurred.

  — Prevent the spread of contamination by covering the spill with absorbent paper. Paper should be dampened if solids are spilled.

  — Clean up the spill, wearing disposable gloves and using absorbent paper.

  — Carefully fold the absorbent paper with the clean side out and place in a plastic bag for transfer to a radioactive waste container. Put contaminated gloves and any other contaminated disposable material in the bag.

— Resurvey the area. Check the area around the spill for contamination. Also check hands, clothing, and shoes for contamination.

— Report the incident to the RSO promptly.

- Reminders to RSO

  — Follow up on the decontamination activities and document the results.

  — As appropriate, determine cause and corrective actions needed; consider bioassays if licensed material may have been ingested or inhaled.

  — If necessary, notify NRC.

Major Spills of Liquids and Solids

- Instructions to Workers

  — Clear the area. If appropriate, survey all persons not involved in the spill and vacate the room.

  — Prevent the spread of contamination by covering the spill with absorbent paper (paper should be dampened if solids are spilled), but do not attempt to clean it up. To prevent the spread of contamination, limit the movement of all personnel who may be contaminated.

  — Shield the source only if it can be done without further contamination or significant increase in radiation exposure.

  — Close the room and secure the area to prevent entry. Post a sign to warn anyone trying to enter that a spill of radioactive material has occurred.

  — Notify the RSO immediately.

  — Survey all personnel who could possibly have been contaminated. Decontaminate personnel by removing contaminated clothing and flushing contaminated skin with lukewarm water and then washing with a mild soap.

  — Allow no one to return to work in the area unless approved by the RSO.

  — Follow the instructions of the RSO (e.g., decontamination techniques, surveys, provision of bioassay samples, requested documentation).

- Reminders to RSO

  — Confirm decontamination of personnel. If decontamination of personnel was not fully successful, consider inducing perspiration by covering the area with plastic. Then wash the affected area again to remove any contamination that was released by the perspiration.

  — Skin contamination must be evaluated to determine potential exposures. Beta-emitting radionuclides have a high potential for resulting in shallow-dose exposures in excess of regulatory limits from small (microcurie) quantities of contamination.

— Supervise decontamination activities and document the results. Documentation should include location and results of surveys and decontamination results.

— Determine root cause and needed corrective actions; consider need for bioassays if licensed material may have been ingested, inhaled, or absorbed.

— If necessary, notify NRC.

Minor Fires

• Instructions to Workers

— If possible, immediately attempt to put out the fire by approved methods (e.g., fire extinguisher) if other fire hazards or radiation hazards are not present.

— Notify all persons present to vacate the area and have one individual immediately call the RSO and fire department or 911 (as instructed by the RSO).

— Once the fire is out, isolate the area to prevent the spread of possible contamination.

— Ensure injured personnel receive medical attention.

— Survey all persons involved in combating the fire for possible contamination.

— Decontaminate personnel by removing contaminated clothing and flushing contaminated skin with lukewarm water, then washing with a mild soap.

— In consultation with the RSO, determine a plan of decontamination and the types of protective devices and survey equipment that will be necessary to decontaminate the area.

— Allow no one to return to work in the area unless approved by the RSO.

— Follow the instructions of the RSO (e.g., decontamination techniques, surveys, provision of bioassay samples, requested documentation).

• Reminders to RSO

— Notify emergency medical personnel of any injured individuals who may be contaminated. Provide radiation safety assistance (e.g., monitoring) as needed or requested.

— Supervise decontamination activities at the facility.

— If decontamination of personnel was not fully successful, consider inducing perspiration by covering the area with plastic. Then wash the affected area again to remove any contamination that was released by the perspiration.

— Consult with fire safety officials to ensure that there is no likelihood of fire restarting and that it is safe to re-enter the building.

— Determine cause and needed corrective actions; consider need for bioassays if licensed material may have been ingested or inhaled. Document the incident.

— If necessary, notify NRC.

Fires, Explosions, or Major Emergencies

- Instructions to Workers

  — Notify all persons in the area to leave immediately.

  — Notify the fire department or 911.

  — Notify the RSO and other facility safety personnel.

  — Ensure injured personnel receive medical attention.

  — Upon arrival of firefighters, inform them where radioactive materials are stored or where radionuclides were being used; inform them of the present location of the licensed material and the best possible entrance route to the radiation area, as well as any precautions to avoid exposure or risk of creating radioactive contamination by use of high pressure water, etc.

  — Allow no one to return to work in the area unless approved by the RSO.

  — Follow the instructions of the RSO (e.g., decontamination techniques, surveys, provision of bioassay samples, requested documentation).

- Reminders to RSO

  — Notify emergency medical personnel of any injured individuals who may be contaminated. Provide radiation safety assistance (e.g., monitoring) as needed or requested.

  — Coordinate activities with local fire department or other emergency personnel.

  — Consult with the firefighting personnel or other emergency personnel and set up a controlled area where personnel can be surveyed for contamination of their protective clothing and equipment after the fire is extinguished.

  — Once the fire is extinguished, provide assistance to firefighters or other emergency personnel who may need to re-enter restricted areas to determine the extent of the damage to the licensed material use and storage areas. To the extent practical, assist firefighters and emergency personnel in maintaining their exposures ALARA if the fire resulted in a significant release of radioactive material or loss of shielding capability, such that excessive radiation levels (greater than 100 millirems per hour) are created.

  — Perform thorough contamination surveys of firefighters and emergency personnel and their equipment before they leave the controlled area, and decontaminate if necessary.

  — Supervise decontamination activities.

  — Consider bioassays if licensed material may have been ingested or inhaled. Document incident.

  — If necessary, notify NRC.

Copies of emergency procedures should be provided to all users. A current copy of the emergency procedures should be posted in each area where radioactive material is used.

APPENDIX R

Model Radiation Survey Procedures

Model Radiation Survey Procedures

This Appendix provides applicants and licensees with additional information on surveys, including training requirements, survey frequency, contamination limits, and bioassays.

Ambient Radiation Level Surveys

- Dose-rate surveys, at a minimum, should be performed in locations where workers are exposed to radiation levels that might result in radiation doses in excess of 10% of the occupational dose limits.

- Dose-rate surveys, at a minimum, should be performed in locations where members of the public could receive a total effective dose equivalent of 1 mSv (100 mrem) in a year, or the dose in any unrestricted area from external sources could exceed 0.02 mSv (2 mrem) in any one hour.

- Dose-rate surveys should be performed in a manner and frequency that is representative of the use of radioactive materials. At a minimum, these surveys should be conducted daily in areas of radioactive material use, where exposures to workers could reasonably occur (e.g., generator storage/elution and dose preparation stations). Other areas, where radiological conditions are not expected to change appreciably from day to day, should be surveyed weekly (e.g., radioactive waste storage areas).

Contamination Surveys

Licensees' contamination surveys should be sufficient to identify areas of contamination that might result in unacceptable levels of exposure to workers or to the public. Combined removable and fixed contamination should be surveyed using appropriate radiation detection equipment. Removable contamination can be detected and measured through wipe tests, which should be analyzed using an appropriate counting instrument. Fixed contamination may be measured directly at the surface of the contamination with the appropriate instrument detector held at close proximity to the surface without direct contact. See Table J.1 for examples of appropriate instruments.

Contamination surveys should be performed:

- To evaluate radioactive contamination that could be present on surfaces of floors, walls, laboratory furniture, or equipment;

- After any spill or contamination event;

- To evaluate contamination of users and the immediate work area at the end of each day when licensed material is used;

APPENDIX R

- In unrestricted areas at frequencies consistent with the types and quantities of materials in use; and

- In areas adjacent to restricted areas and in all areas through which licensed materials are transferred and temporarily stored before shipment.

Contamination Survey Frequency

All areas where radioactive materials are eluted, prepared, assayed, dispensed, or packaged for transport should be surveyed daily. All other areas where radioactive materials are used or stored should be surveyed weekly.

Contamination in Unrestricted Areas

Contamination found in unrestricted areas should be immediately decontaminated to background levels. When it is not possible to get to background levels, the licensee must ensure that the amounts do not exceed the contamination levels listed in Table R.1.

Table R.1   Recommended Action Levels in dpm/100 cm² for Removable Surface Contamination by Radiopharmaceuticals

	C-11, N-13, O-15, F-18, P-32, Se-75, Sr-85, Sr-89, In-111, I-123, I-125, I-131, Sm-153, Yb-169, Re-186, Au-198	Cr-51, Ga-67, Tc-99m, Tl-201
1. Unrestricted areas, personal clothing	200	2000
2. Restricted areas, protective clothing used only in restricted areas, skin	2000	20000

When equipment or facilities that are potentially contaminated are to be released for unrestricted use, the above table provides the maximum acceptable residual levels. To the extent practicable, it is appropriate to decontaminate below these levels. Surface contamination surveys should be conducted for both removable and fixed contamination before these facilities or equipment are released from restricted to unrestricted use to ensure that they meet these limits.

A standardized method for wipe testing of a relatively uniform area should be used to aid in comparing contamination at different times and places. A wipe taken from an area of approximately 100 cm² is acceptable to indicate levels of removable contamination.

Survey Record Requirements

Each survey report should include the following:

• Diagram of the area identifying specific locations surveyed (See Figure 8.3),

• Ambient radiation levels with appropriate units,

• Contamination levels with appropriate units,

• Make and model number of instruments used,

• Background levels,

• Name of the person making the evaluation and recording the results and date, and

• Corrective actions taken for elevated levels identified and results of resurveys.

Licensees should record contamination levels observed and procedures followed for incidents involving contamination of individuals. The record should include names of individuals involved, description of work activities, calculated dose, probable causes (including root causes), steps taken to reduce recurrence of contamination, times and dates, and surveyor's signature.

Air Sampling

Air sampling can be used to do the following:

• Determine whether the confinement of radioactive materials is effective,

• Measure airborne radioactive material concentrations in the workplace,

• Estimate worker intakes of radioactive material,

• Determine posting requirements,

• Determine what protective equipment and measures are appropriate, and

• Warn of significantly elevated levels of airborne radioactive materials.

Refer to Regulatory Guide 8.25, Revision 1, "Air Sampling in the Workplace," dated June 1992, and NUREG-1400, "Air Sampling in the Workplace," dated September 1993, for further guidance on air sampling.

Air Stack Release Monitoring

Airborne radioactive effluents should be monitored at the release points (e.g., stack) to provide accurate measurements to estimate public exposure. Licensees should verify the performance of effluent monitoring systems by regular calibration of equipment and checks of filtration to ensure their reliability.

Regulatory Guide 4.20, "Constraints on Release of Airborne Radioactive Materials to the Environment for Licensees Other Than Power Reactors," dated December 1996, provides guidance on methods acceptable (calculation or COMPLY code) to NRC for compliance with the constraint on air emissions to the environment.

Regulatory Guide 8.37, "ALARA Levels for Effluents from Materials Facilities," dated July 1993, provides guidance on designing an acceptable program for establishing and maintaining ALARA levels for gaseous and liquid effluents at materials facilities.

Effluent monitoring systems should be designed in accordance with ANSI N13.1 (1999), "Sampling and Monitoring Releases of Airborne Radioactive Substances From the Stacks and Ducts of Nuclear Facilities," and ANSI N42.18, "Specification and Performance of On-Site Instrumentation for Continuously Monitoring Radioactivity in Effluents."

Radioiodine Monitoring

The handling of radioiodine requires additional surveys and monitoring. Such surveys and monitoring include:

- Routine surveys should be performed of air filters incorporated in fume hoods and gloveboxes to identify when filters should be exchanged prior to saturation.

- Routine surveys should be performed in the area where radioiodine is handled immediately following each use to identify elevated radiation and contamination levels.

- Continuous monitoring of the air effluent should be performed during radioiodine use. In-line filters should be monitored periodically to determine actual effluents.

Sanitary Sewerage Release Monitoring

The licensee should evaluate the concentrations of radioactive material in water that is released to the environment and to the sanitary sewer. The licensee must show that these releases meet the limits in 10 CFR 20. 1301 and 20.2003, respectively.

Bioassay Monitoring

Frequency of Required Bioassay Measurements

Determining the appropriate frequency of routine bioassay measurements depends upon the exposure potential and the physical and chemical characteristics of the radioactive material and the route of entry to the body. Consider the following elements:

- Potential exposure of the individual,

- Retention and excretion characteristics of the radionuclide,

- Sensitivity of the measurement technique, and

- Acceptable uncertainty in the estimate of intake and committed dose equivalent.

Bioassay measurements used for demonstrating compliance with the occupational dose limits should be conducted often enough to identify and quantify potential exposures and resultant intakes that, during any year, are likely to collectively exceed 0.1 times the annual limit on intake (ALI). The 10% ALI criterion is consistent with 10 CFR 20.1502(b), which requires licensees to monitor intakes and assess occupational doses for exposed individuals who are likely to exceed 10% of the applicable limit (i.e., intakes likely to exceed 0.1 ALI for adults).

Separate categories of bioassay measurements, routine measurements, and special measurements further determine the frequency and scope of measurements.

Routine Measurements

Routine measurements include baseline measurements, periodic measurements, and termination measurements. These measurements should be conducted to confirm that appropriate controls exist and to assess dose.

An individual's baseline measurement of radioactive material within the body should be conducted before beginning work that involves exposure to radiation or radioactive materials for which monitoring is required.

In addition to the baseline measurements, periodic bioassay measurements should be performed. The frequency of periodic measurements should be based on the likelihood of significant exposure of the individual. In determining the worker's likely exposure, consider such information as the worker's access, work practices, measured levels of airborne radioactive material, and exposure time. Periodic measurements should be made when the cumulative exposure to airborne radioactivity, since the most recent bioassay measurement, is > 0.02 ALI (40 derived air concentration (DAC) hours). Noble gases and airborne particulates with a radioactive half-life of less than two hours should be excluded from the evaluation, since external exposure generally controls these radionuclides.

When an individual is no longer subject to the bioassay program because of change in employment status, termination bioassay measurements should be made, when practicable, to ensure that any unknown intakes are quantified.

Special Monitoring

Because of uncertainty in the time of intakes and the absence of other data related to the exposure (e.g., physical and chemical forms, exposure duration), correlating positive results to actual intakes for routine measurements can sometimes be difficult. Abnormal and inadvertent intakes from situations such as inadequate engineering controls, inadvertent ingestion, contamination of a wound, or skin absorption, should be evaluated on a case-by-case basis. When determining whether potential intakes should be evaluated, consider the following circumstances:

- Presence of unusually high levels of facial and/or nasal contamination,

- Operational events with a reasonable likelihood that a worker was exposed to unknown quantities of airborne radioactive material (e.g., loss of system or container integrity),

- Known or suspected incidents of a worker ingesting radioactive material, and

- Incidents that result in contamination of wounds or other skin absorption.

References

1. Regulatory Guide 4.20, "Constraints on Release of Airborne Radioactive Materials to the Environment for Licensees Other Than Power Reactors," dated December 1996.

2. Regulatory Guide 8.9, Revision 1, "Acceptable Concepts, Models, Equations, and Assumptions for a Bioassay Program," dated July 1993.

3. Regulatory Guide 8.25, Revision 1, "Air Sampling in the Workplace," dated June 1992.

4. Regulatory Guide 8.37, " ALARA Levels for Effluents from Materials Facilities," dated July 1993.

5. NUREG-1400, "Air Sampling in the Workplace," dated September 1993.

6. NUREG/CR - 4884, "Interpretation of Bioassay Measurements," dated July 1987.

7. ANSI N13.1, "Sampling and Monitoring Releases of Airborne Radioactive Substances From the Stacks and Ducts of Nuclear Facilities," dated 1999.

8. ANSI N42.18, "Specification and Performance of On-site Instrumentation for Continuously Monitoring Radioactivity in Effluents," dated 2004.

APPENDIX S

Model Procedure for Return of Radioactive Wastes from Customers

Model Procedure for Return of Radioactive Wastes from Customers

Procedures for Customers to Return Radioactive Waste to the Radiopharmacy

Return only items that contained or contain radioactive materials supplied by the radiopharmacy (e.g., pharmacy-supplied syringes and vials and their contents). Most return shipments to radiopharmacies will qualify as excepted packages of limited quantity, in accordance with DOT requirements (49 CFR 173.421). For those packages containing radioactive material in excess of the limited quantity, customers should ensure that all applicable DOT requirements are met for the packages. These include, but are not limited to, certification packaging (Type A), package marking and labeling, and shipping papers. For specific guidance on preparing these types of packages, follow the in-house procedures for shipping radioactive material packages or contact the pharmacy for guidance.

Preparation of radioactive materials for return as an excepted package of limited quantity:

- Ensure that the activities of material being returned are limited quantities as defined by DOT in Table 4 of 49 CFR 173.425. Special attention should be given for the return of unused doses that may still contain significant activities of radionuclides. The amount of radioactivity in unused doses may necessitate that a syringe or vial be held for decay to reduce the activity to that permitted for shipment of limited quantities.

- Place the syringe or vial in the original, labeled, lead shield in which it was delivered; and

- Place shielded waste into the shipping package (e.g., padded briefcase or ammo box) in which it was delivered. Note: Packages used to ship radioactive material to customers meet the DOT package requirements for transport of limited quantities.

Preparation of package:

- Using a calibrated survey meter, measure the radiation levels at all points on the surface of the package to ensure that levels are less than or equal to 0.5 mrem/hr;

- Use contamination wipes on the surface of the package to ensure that the removable contamination does not exceed the limit specified in 49 CFR 173.443(a), 22 dpm/cm^2 over a 300 cm^2 area;

- Label the package as an "Excepted Package - Limited Quantity of Material;" and

- Seal the package so that it will be evident upon receipt if the package was accidentally opened during shipment.

Procedure for Receipt and Opening of Packages from Customers Containing Radioactive Waste

- Place all returned packages in an identifiable location within the radiopharmacy.

- Put on disposable gloves.

- Monitor the package for removable contamination. If wipe tests indicate contamination levels greater than 22 dpm/cm^2 over a 300 cm^2 area: notify the customer and NRC, survey the driver/courier who retrieved the waste and the vehicle used to transport the waste to the radiopharmacy, and decontaminate the package or remove it from service for decay.

- Open the package and identify each nuclide in the shielded containers.

- Dispose of radioactive waste into the appropriate container for the half-life of the nuclide being disposed of, in accordance with the radiopharmacy's procedures for disposal of waste by decay-in-storage.

- Survey the transport radiation shields for contamination with a low-level survey meter. Any transport radiation shield that indicates activity exceeding background readings should be decontaminated or removed from service.

APPENDIX T

NRC Incident Notifications

NRC Incident Notifications

Table T.1 Typical Notifications Required for Radiopharmacy Licensees

Event	Telephone Notification	Written Report	Regulatory Requirement
Theft or loss of material	immediate	30 days	10 CFR 20.2201(a)(1)(i)
Whole body dose greater than 0.25 Sv (25 rems)	immediate	30 days	10 CFR 20.2202(a)(1)(i)
Extremity dose greater than 2.5 Sv (250 rems)	immediate	30 days	10 CFR 20.2202(a)(1)(iii)
Intake of five times the annual limit on intake	immediate	30 days	10 CFR 20.2202(a)(2)
Removable contamination exceeding the limits of 10 CFR 71.87(i) - (beta/gamma/low toxicity alpha - 22 dpm/cm^2; all other alpha - 2.2 dpm/cm2)	immediate	none	10 CFR 20.1906(d)(1)
External radiation levels exceeding the limits of 10 CFR 71.47 - (any point on the surface - 2 mSv/hr (200 mrem/hr))	immediate	none	10 CFR 20.1906(d)(2)
Whole body dose greater than 0.05 Sv (5 rems) in 24 hours	24 hours	30 days	10 CFR 20.2202(b)(1)(i)
Extremity dose greater than 0.5 Sv (50 rems) in 24 hours	24 hours	30 days	10 CFR 20.2202(b)(1)(iii)
Intake of one annual limit on intake	24 hours	30 days	10 CFR 20.2202(b)(2)
Occupational dose greater than the applicable limit in 10 CFR 20.1201	none	30 days	10 CFR 20.2203(a)(2)(i)
Dose to individual member of public greater than 1 mSv (100 mrems)	none	30 days	10 CFR 20.2203(a)(2)(iv)
Defect in equipment that could create a substantial safety hazard	2 days	30 days	10 CFR 21.21(d)(3)(i)
Filing petition for bankruptcy under 11 U.S.C.	none	immediately after filing petition	10 CFR 30.34(h)

Event	Telephone Notification	Written Report	Regulatory Requirement
Expiration of license	none	60 days	10 CFR 30.36(d)
Decision to permanently cease licensed activities at *entire site*	none	60 days	10 CFR 30.36(d)
Decision to permanently cease licensed activities in any *separate building or outdoor area* that is unsuitable for release for unrestricted use	none	60 days	10 CFR 30.36(d)
No principal activities conducted for 24 months *at the entire site*	none	60 days	10 CFR 30.36(d)
No principal activities conducted for 24 months *in any separate building or outdoor area* that is unsuitable for release for unrestricted use	none	60 days	10 CFR 30.36(d)
Event that prevents immediate protective actions necessary to avoid exposure to radioactive materials that could exceed regulatory limits	immediate	30 days	10 CFR 30.50(a)
An unplanned contamination event involving greater than 5 times the ALI, and half-life greater than 24 hours requiring that access be restricted for more than 24 hours	24 hours	30 days	10 CFR 30.50(b)(1)
Equipment is disabled or fails to function as designed when required to prevent radiation exposure in excess of regulatory limits	24 hours	30 days	10 CFR 30.50(b)(2)
Unplanned fire or explosion that affects the integrity of any licensed material or device, container, or equipment with licensed material	24 hours	30 days	10 CFR 30.50(b)(4)

Note: Telephone notifications shall be made to the NRC Operations Center, at 301-816-5100 or 301-951-0550.

APPENDIX U

Summary of Comments Received on Draft NUREG-1556, Volume 13

Summary of Comments Received on NUREG-1556, Volume 13

Table U.1   Comment from Jose O. Morales, M.D., Dated March 19, 1999

Location	Subject	Comment
Appendix S	Return Shipments to Radiopharmacies	Appendix S of NUREG-1556, Vol. 13 does not allow for the return of unused doses until they have decayed enough to qualify for shipment as a "limited quantity." This will create additional problems, requiring the need for additional space in the "hot lab." The present practice of returning unused doses works well since it is the same carrier that takes back the material. The preparation of the package is done following the present standards. Therefore, I suggest that both options be allowed.

NRC Staff Response: It was not the writing team's intent to limit return shipments to radiopharmacies to excepted packages of limited quantity. Appendix S has been modified to clarify that return shipments are not limited to excepted packages. Since most return shipments will likely be as excepted packages of limited quantity, we have elected to limit the guidance in the NUREG to these types of packages.

**Table U.2   Comment from Carol S. Marcus, Ph.D., M.D.,
Dated December 21, 1998**

Location	Subject	Comment
Appendix R, Table R.1	Removable Contamination Survey Action Limits	I read with interest NRC's "acceptable license termination screening values of common radionuclides for building surface contamination," which appear in Table 1 of Fed. Reg. 63 (222) 64134, attached. These actually appear to have been based on science. Compare these with NRC's "acceptable surface contamination levels in unrestricted areas," found in Table R.3, p. R-4 of Draft "Program-Specific Guidance About Medical Use Licensees", NUREG-1556 Vol. 9, attached. (The comment was also applicable to NUREG-1556, Vol. 13) *For the same radionuclide,* the medical licensee limits are about 3-10,000 times more restrictive than for decommissioned licensees. Please explain.

NRC Staff Response: The contamination levels listed in Table 1 of the Federal Register notice are for relatively long-lived radionuclides that are likely to result in 25 millirem per year to members of the public. Therefore, use of the Table and the dose criteria do not lend themselves to meaningful use for those licensees who use short half-life materials, such as radiopharmacies. The values in Appendix R, Table R.1 of the Draft NUREG, are suggested values that have historically been used in other NRC guidance documents and been demonstrated to be reasonably achievable. Applicants and licensees are able to set their own values for contamination limits in unrestricted areas, as long as the values are as low as is reasonably achievable. No changes have been made to the Draft guidance.

Table U.3 Comments from Mallinckrodt, Inc., Dated June 7, 1999

Location	Subject	Comment
Section 8.2	Timely Notification of Transfer of Control	The criteria states that licensees must provide full information and provide NRC's *prior written consent* before transferring control of the license, or, as some licensees call it, "transferring the license." This criteria may be difficult for the licensees to implement - particularly since the details of NRC's requirements are not generally available to the licensee in advance. This requirement may result in an unnecessary licensing burden. Instead, we suggest that the criteria for advance notice be better defined and those that are likely going to be met - for example, unlicensed possessions or lack of management oversight.

NRC Staff Response: The NUREG provides guidance on the type of information required by NRC in order to evaluate potential control transfers. Section 30.34 of 10 CFR Part 30 specifically prohibits the transfer of a license, either voluntarily or involuntarily, unless the NRC finds that the transfer is in accordance with the provisions of the Act and gives its consent in writing. The guidance includes the type of information required by NRC for its review and this information has been included in Appendix F of the NUREG. This is so stated in the Response from Applicant section on the same page (8-3); therefore, no additional guidance or changes are necessary to further address this comment.

Location	Subject	Comment
Section 8.5.1	Manipulation of Volatile Materials	With respect to the use of potentially volatile materials (e.g., 1-131), the regulatory guide requires for the applicant to specify whether the material will be manipulated at the radiopharmacy in a volatile form. A clarification is warranted between the terms "manipulation" and "compounding." In particular, it should be clarified if the handling of sealed I-131 vials for the purpose of re-distribution is classified as a "manipulation" - as compared to handling of unsealed I-131 under a glovebox or a fume hood for the purpose of manipulating the desired therapeutic activity of liquid I-131 or producing the desired activity in a capsule form.

NRC Staff Response: The guidance has been clarified to indicate that "manipulation" does not apply to the re-distribution of sealed I-131 vials, as long as the seal remains intact.

Location	Subject	Comment
Section 8.5.1	Use of Sealed Sources	The regulatory guide states that the applicants will be authorized to possess and use only those sealed sources, such as calibration and reference sources, that are specifically approved or registered by NRC or an Agreement State. NRC is going to require the applicants to provide the manufacturer's name and model number for each requested sealed source and device so that NRC can verify that they have been evaluated in an SSD Registration Certificate or specifically approved on a license. We suggest, instead, that the applicants should be able to provide to NRC reference of the approved SSD Registrants with a specific upper activity limit for each sealed/reference source and without the make and model number. This will allow the flexibility for the radiopharmacies to procure the desired sources on an as needed basis - as long as the possession limits are not exceeded. This current practice appears to be functioning effectively, and should not be changed.

NRC Staff Response: Under 10 CFR 30.32(g), an application for a specific license to use byproduct material in the form of a sealed source or in a device that contains the sealed source must either (1) identify the source or device by manufacturer and model number as registered with the NRC under 10 CFR 32.210 or with an Agreement State; or (2) contain the information identified in 10 CFR 32.210. Applicants may gain greater flexibility by identifying any sources or devices they may possess by manufacturer and model number and by requesting a possession limit sufficient to cover their use.

Location	Subject	Comment
Section 8.5.1	Use of Depleted Uranium	Please note that the DU shielding provided for the Mo-99/Tc-99m generators qualifies for the exemption under 10 CFR 40.13(c)(6). Therefore, the DU shielding information required for submittal by the applicants should be deleted.

NRC Staff Response: NRC clarified its position with regard to depleted uranium (DU) shielding used in generators in Policy and Guidance Directive (P&GD) 86-9, "Authorizing Possession and Use of Depleted Uranium as Shielding for High Activity Molybdenum-99/Technetium-99m Generators," dated June 1986. This P&GD will be considered superseded with the issuance of NUREG-1556, Vol. 13, in final form and, therefore, the general guidance contained in that document was incorporated into the Draft NUREG. The P&GD states, in summary, that:

"Depleted uranium associated with Mo-99/Tc-99m generators is exempted from licensing requirements (see 10 CFR 40.13(c)(6)) only when it is used as a shipping container (e.g., when the generator is in transit from its manufacturer to the pharmacy). However, a specific license or authorization from NRC is needed to possess and use the depleted uranium as a shield (e.g., during the time that the pharmacy uses or stores the Mo-99/Tc-99m generator at its facility)."

Therefore, applicants that intend to use DU as shielding must provide the requested information in order to be authorized.

Location	Subject	Comment
Section 8.5.1	Activity Levels of Generators Using DU as a Shield	It is not necessarily true that the DU shielding is frequently used for generators with Mo-99 activity in excess of 148 GBq (4 Curies). Some licensees may not use the DU shielding unless the activity is even higher than 4 Curies. We suggest that the reference to any specific activity level be deleted.

NRC Staff Response: The activity level referenced is accurate. Therefore, no change has been made.

Location	Subject	Comment
Section 8.5.2	Financial Assurance	The Discussion section of this item indicates that this requirement is not applicable to the commercial radiopharmacies since the vast majority of radioactive materials they possess and re-distribute do not have half-lives greater than 120 days. Please note, however, that radiopharmacies may possess sealed sources with greater than 120-day half-lives. The above statement should be clarified to indicate that the sealed sources may be excluded from the financial requirement.

NRC Staff Response: The guidance already addresses this issue by referencing the activity level thresholds referenced in 10 CFR 30.35(b) and (d), which include financial assurance requirement thresholds for sealed sources.

Location	Subject	Comment
Section 8.6.1	Re-distribution of Sealed Sources	The regulatory guide requires an applicant's response to include a confirmation that the manufacturer's labeling and packaging will not be altered for re-distribution of sources (e.g., to radiopharmacy customers). We believe that this requirement creates an unnecessary licensing burden on the radiopharmacies, without any added benefit to health and safety. As long as the redistribution of these sources is performed in accordance with the performance-based NRC and DOT regulatory requirement, the prescriptive guidance provided in this section is unwarranted and should be deleted. Please note that the primary packaging provided by the manufacturer is frequently altered since the radiopharmacies generally use their own packaging for transporting orders to their customers.

NRC Staff Response: Section 10 CFR 32.74(a)(3) requires that applicants for licenses to manufacture and distribute sources or devices containing byproduct material for medical use must describe the label affixed to the source or device or to the permanent storage container for the source or device. The label must contain information on the radionuclide, quantity, and date of assay, and a statement that the U.S. Nuclear Regulatory Commission has approved distribution of the (name of source or device) to persons licensed to use byproduct material identified in Sections 35.57, 35.400, and 35.500 of 10 CFR Part 35, as appropriate, and to persons who hold an equivalent license issued by an Agreement State. If other persons wish to redistribute sources or devices previously approved for distribution under 10 CFR 32.74, the person redistributing may not alter, remove, cover, or deface the label affixed by the initial distributor to meet the requirements of 10 CFR 32.74(a)(3). If applicants wish to affix their own label to sources or devices for redistribution to persons licensed to use byproduct material identified in Sections 35.57, 35.400, and 35.500 of 10 CFR Part 35, and to persons who hold an equivalent license issued by an Agreement State, the applicant must apply for and receive specific authorization pursuant to 10 CFR 32.74. To the extent that the manufacturer's original packaging is an integral part of its authorization to distribute the source or device in accordance with 10 CFR 32.74, persons who desire to redistribute the source or device must use that original packaging; however, if the packaging is not specified in the approval for initial distribution, other persons may repackage the source or device for redistribution.

Location	Subject	Comment
Section 8.7.2	State Registration of Authorized Nuclear Pharmacists	The response from an applicant requires, among other things, a copy of the state pharmacy license or registration for the pharmacist. Please note that the radiopharmacy licenses issued to Mallinckrodt in various states includes the names of approximately 100 Authorized Nuclear Pharmacists (ANPs) - these names were included in one of our licenses prior to December 2, 1994. Subsequently, the remaining radiopharmacy licenses issued to Mallinckrodt pharmacies were amended to name the same ANPs. We are assuming that, upon renewal of the existing pharmacy licenses and for new pharmacy license applications, we will be able to continue to reference the existing ANPs, even though the pharmacy licenses are issued to these ANPs by different states.

NRC Staff Response: NRC's definition of "Authorized Nuclear Pharmacist" does not require the individual to be registered in the State in which he or she practices. The individual must be a pharmacist and demonstration of this is usually accomplished by evidence of State licensure or registration. Although an individual may qualify as an ANP under NRC's regulations, nothing in that qualification relieves the licensee or the ANP from complying with other applicable Federal or State requirements governing radioactive drugs.

Location	Subject	Comment
Section 8.9	Description of Ventilation Systems	The response required by the applicants includes the description of ventilation systems, including gloveboxes or fume hoods, with pertinent airflow rates, ventilation systems, and monitoring systems. While we commend NRC for proposing the performance-based approach to the licensing of radiopharmacies, we believe that the above requirements are rather too prescriptive, and not in the spirit of the performance-based rulemaking. It is likely that, under the requirements of the present regulatory guide, any changes in the ventilation system may require an amendment to the license. We suggest that NRC provide only performance-based regulatory requirements to the applicants - in terms of the facilities and equipment.

NRC Staff Response: The applicant is not required to provide descriptions of specific systems and operating parameters but may include a description of minimal performance and operating criteria.

Location	Subject	Comment
Section 8.10.1	Licensee Self-identified Deficiencies	This regulatory guide indicates that NRC will review the licensee's audit results and determine if corrective actions are thorough, timely, and sufficient to prevent recurrence. If violations are identified by the licensee and these steps are taken, NRC can exercise discretion and will normally elect not to cite a violation. As indicated in the report, NRC's goal is to encourage prompt identification and prompt comprehensive correction of violations and deficiencies. It is imperative that NRC recognize the efforts of a licensee to identify and take appropriate actions for their "self-identified" deficiencies and not to penalize the licensee for its pro-active regulatory compliance program. Therefore, the effectiveness of a program should be evaluated based on the end results rather than the contents of the internal audit reports.

NRC Staff Response: In order to evaluate the effectiveness of a licensee's self-assessment and corrective action program, NRC must review the nature of the licensee's findings during program audits and evaluate the appropriateness of the proposed or enacted corrective actions. It is not NRC's intention to "penalize" licensees for proactive regulatory compliance programs. For this reason, NRC's enforcement policy (NUREG-1600) specifically affords inspectors the authority to withhold the issuance of a Notice of Violation for licensee-identified violations in those cases where it is warranted and appropriate.

Location	Subject	Comment
Section 8.10.3, Table 8.1	Type B Packaging	Some of the requirements in Table 8-1 are related to Type B packages (packages containing quantities greater than Type A). Since radiopharmacies do not generally either receive or ship Type B packages, this information should be deleted.

NRC Staff Response: The NRC recognizes that radiopharmacies do not generally receive or ship these types of packages; however, the guidance is provided for those licensees and applicants for whom it may be warranted.

Location	Subject	Comment
Section 8.10.8	Dose Calibrator Beta-Correction Factors	The discussion and the applicant's response require a sample calculation for determining beta-correction for dose calibrators with ionization chambers for beta emitting radionuclides. However, 10 CFR 35.53(b) provides an exemption for unit doses prepared by manufacturers or prepared under 10 CFR 32.72 (or equivalent Agreement State requirements). This information should be added under Item 10.8.8. (sic)

NRC Staff Response: Since 10 CFR 35.53(b) does not apply to licensed radiopharmacies, discussion of the exemption described in the comment is not appropriate for this guidance document. If radiopharmacy applicants intend to only redistribute beta-emitting radionuclides that have been previously prepared and distributed by other persons licensed pursuant to 10 CFR 32.72, then the correction factor calculation is not required. If radiopharmacies intend to initially distribute; i.e., measure, prepare, and label, beta-emitting radionuclides, the applicant must provide the calculation to demonstrate its ability to accurately dispense such materials.

Location	Subject	Comment
Section 8.10.12	Transport Radiation Shields	The regulatory guide provides, as general guidelines, the surface dose rates for the "transport radiation shields" (e.g., not more than 3 mrem/hr of Tc-99m products, 50 mrem/hr for diagnostic dosages of I-131, and up to 150 mrem/hr for therapeutic I-131 dosages). As stated earlier in our comments under Comment #6 (Item 8.9 (Item 9), Facilities and Equipment), we believe that the above dose rate criteria for the transport radiation shields are rather prescriptive and do not fall under the performance-based approach of the requirements of this regulatory guide. We suggest that these dose rate criteria be deleted, and, instead, the licensees should be evaluated based on compliance with the DOT requirements for transporting radiopharmaceutical products. We agree, however, that the applicant should select appropriate shielding materials and dimensions to not only ensure that the occupational doses are ALARA, but also that the "transport radiation shields" can be easily handled for occupational safety.

NRC Staff Response: The dose rates contained in the "Discussion" Section of the guidance are provided as a reference to dose rate emissions for previously NRC-approved transport radiation shields. The dose rates are not prescriptive, but informative. The decision to approve or deny new transport radiation shields will be based on the ability of the shield to achieve compliance with DOT requirements as well as to maintain worker (pharmacy and customer) radiation exposures as low as is reasonably achievable, while assuring ease of handling for occupational safety purposes.

Location	Subject	Comment
Appendix I	Dose Calibrator Testing Deviations	The audit checklist under Equipment and Instrumentation (G-Dose Calibrators for Photon-Emitters, and H-Dose Measurement Systems for Beta and Alpha Emitters, 10 CFR 32.72©), the constancy, linearity, geometry, and accuracy deviations are indicated as + 10%. We believe the deviations should be indicated as ± 10% (as correctly indicated in Appendix O).

NRC Staff Response: Comment acknowledged and correction made.

Location	Subject	Comment
Appendix N	Conversion	Item 2 under the training program indicates the training criteria for individuals who are likely to receive an occupational dose in excess of 100 mSv (100 mrem). Please note that the criteria should be corrected to indicate 1 mSv (100 mrem).

NRC Staff Response: Comment acknowledged and correction made.

Location	Subject	Comment
Appendix O, Item 2	Installation of Dose Calibrators	Model program requires repeat of the above tests (constancy, linearity, geometry, accuracy) after repair, adjustment, or relocation to another building. The present NRC interpretation indicated to Mallinckrodt radiopharmacies is that these tests are required after "any" relocation. Please clarify the interpretation of this requirement - in particular, is the relocation intended to mean only upon relocation to another building?

NRC Staff Response: Operational tests must be repeated when the dose calibrator is relocated to another location, if the relocation involves extensive handling. If the handling could reasonably call into question the proper functioning of the ionization chamber or electronics, then the applicable tests must be repeated. Relocation includes movement to another location within the same building.

Location	Subject	Comment
Appendix O, Item 4 and 4.2.5.1	Lower Limit of Linearity Testing	There appears to be a conflict in these two requirements; Item 4 indicates that the linearity of a dose calibrator should be ascertained over the range of its use between the maximum activity in a vial and 30 microcuries, whereas 4.2.5.1 indicates the lower ranges as 10 microcuries. Please provide the corrected values for the lower range. We believe the correct value should be 30 microcuries, as specified in 10 CFR 35.50(b)(3).

NRC Staff Response: The range in 4.2.5.1 should be changed to indicate a lower range of 30 microcuries for linearity testing.

Location	Subject	Comment
Appendix O, Item 1.2 and 4.2.5.5	Acceptable Linearity Deviation	There appears to be a conflict in these two requirements: Item 1.2 indicates a deviation of ±10% whereas Item 4.2.5.5 indicates a deviation of 5% for the linearity test. Please make the necessary corrections to indicate that the acceptable deviation is ±10%.

NRC Staff Response: Item 1.2 states that an acceptable level of deviation is ±10%. Item 4.2.5.5 indicates that if the deviation is more than ±5%, then the dose calibrator should be repaired or replaced. If not repaired or replaced, then it is necessary to make a correction table or graph that will allow conversion from activity indicated by the dose calibrator to "true activity." There is no conflict between these two sections. The maximum acceptable error is 10%. For errors between 5 and 10%, the dose calibrator reading must be corrected to reflect the "true activity" of the measured dose.

Table U.4 Comments from the Council on Radionuclides and Radiopharmaceuticals, Inc. (CORAR), Dated June 2, 1999

Location	Subject	Comment
Appendix C and Appendix D	Licensee Versus NRC Model Formats	There are two listings for Appendix D. Appendix C and the second Appendix D both have the same title, "Suggested Format for Providing Information Requested in Items 5 through 11 on NRC Form 313." This is confusing. Whereas this title is appropriate for Appendix C, the Appendix D is to be used as a checklist by NRC to review applications and the second listing for Appendix D should be re-titled to reflect this use.

NRC Staff Response: The title of Appendix D has been changed to reflect its use as a checklist for NRC license reviewers.

Location	Subject	Comment
Section 8.2	Timely Notification of Transfer of Control	Page 8-3 and Appendix F both state that licensees must provide full information and obtain *NRC's prior written consent* before transferring control of the license. Our experience tells us that this is an unreasonable expectation of NRC and one that is subject to inconsistent interpretation at the regional level. In some cases, NRC has cited licensees for failing to obtain prior written approval when there have been changes such as changes in subsidiary business relationships or name changes of subsidiaries where the effect of day-to-day control of and responsibility for licensed activities has not changed. In other cases, similar changes have been reported informally to NRC before and after the fact without formal written consent required.

Regardless of the inconsistent approach taken by NRC on this, the difficulty we have as licensees is that the details required by NRC concerning these activities are rarely available in advance. The expectation is one that cannot be achieved and sets up licensees for needless noncompliance. |

NRC Staff Response: This comment has already been addressed in the response to the comment from Mallinckrodt, Inc., which is similar.

Location	Subject	Comment
Section 8.2	Timely Notification of Transfer of Control	In the third paragraph on page 8-7 it states that applicants should review requirements for financial surety arrangements for decommissioning before specifying possession limits of any radioisotope with a half-life greater than 120 days. This statement should address the fact that material such as this possessed in the form of sealed sources may be excluded from financial assurance requirements.

NRC Staff Response: This comment has already been addressed in the response to the comment from Mallinckrodt, Inc., which is similar.

Location	Subject	Comment
Section 8.5.1	Sealed Versus Unsealed Materials	In the discussion of the contents of license application, the terms sealed and unsealed are used in a fashion that is misleading. These terms need to be more clearly defined and the fact that material in a form other than sealed sources can be contained and not in an open form should be considered. If "sealed" means "sealed sources," then use of the term "unsealed" as in the beginning of the second paragraph on page 8-7 under 8.5.1 is inappropriate. A sealed vial of radioiodine does not present the same concerns over volatility as does open form material removed from a vial. In many licenses, the form of material for which possession is authorized is "any" unless the form is specified as "sealed source." It would be useful to limit the forms authorized in licenses to "any" and "sealed source" while providing some degree of credit in this guide for containment of material to sealed forms other than those encapsulated as "sealed sources."

NRC Staff Response: The NRC has consistently applied the terms "sealed" and "unsealed" sources for all of its licensees, regardless of particular nuances used by specific professions. As used in this, and all NRC references, "sealed source" means any byproduct material that is encased in a capsule designed to prevent leakage or escape of the byproduct material. This is NRC's definition of "sealed source," and can be found in 10 CFR 30.5. The design of sealed sources is such that access to the encased byproduct material can only be gained by destruction of the containing capsule. Unsealed sources are, by default, everything else that does not meet the definition of a sealed source.

Location	Subject	Comment
Section 8.5.1	Use of Depleted Uranium	Possession of depleted uranium as shielding is exempt under 10 CFR 40.13(c)(6). The shielding provided in Mo-99 generators qualifies for this exemption. The instruction provided and the reference to the use of depleted uranium in Mo-99 generators on page 8-8, on page C-4 of Appendix C, under 8.6.2 on page 8-15 and in the table on page D-4 of Appendix D should be removed.

NRC Staff Response: This comment has already been addressed in the response to the comment from Mallinckrodt, Inc., which is similar.

Location	Subject	Comment
Section 8.5.1	Manipulation of Volatile Materials	Under 8.5.1 on page 8-8, the instruction states that the applicant must specify whether potentially volatile material will be "manipulated" at the pharmacy. The term "manipulated" needs to be defined in order for an applicant to provide the appropriate response. In other words, "manipulated" for an applicant may be interpreted as including the handling of sealed vials of radioiodine for the purposes of redistribution to customers. It could also be the handling of open form radioiodine in a hood to produce therapeutic capsules for distribution. The use of the term is unclear and this needs to be addressed to prevent applicants from providing needlessly specific detail.

NRC Staff Response: This comment has already been addressed in the response to the comment from Mallinckrodt, Inc., which is similar.

Location	Subject	Comment
Section 8.6.1	Compounding Non-FDA-Approved Radiopharmaceuticals	There is a lengthy and comprehensive discussion under 8.6.1 on page 8-12 concerning the distribution and redistribution of medical use material. This characterization should include the compounding of non-FDA-approved radiochemicals by an applicant or pharmacy currently licensed by NRC or Agreement States for the *redistribution of radiopharmaceuticals* to customers licensed by NRC or Agreement States to possess radiopharmaceuticals. In addition to including this scenario in the discussion, NRC should also state its position as to whether or not this is acceptable from a licensing standpoint. NRC's position should then be considered in the response instructions provided on page 8-13 and under 8.6.2 on page 8-15 for radiopharmaceuticals.

NRC Staff Response: The NRC regulates the possession, use, and distribution of radiochemicals and radiopharmaceuticals from the standpoint of worker and public radiation protection. The fitness of a particular radiochemical for use in compounding radiopharmaceuticals for ultimate use in medicine is outside NRC's regulatory authority, and therefore, discussion of this issue is not appropriate in this guidance document.

Location	Subject	Comment
Section 8.6.1	Redistribution of Sealed Sources	On pages 8-13 and 8-14, and on pages C-1 and C-3 of Appendix C, it is stated that the applicant's response must provide a confirmation that the manufacturer's packaging, labeling and shielding will not be altered when sources are redistributed. This instruction applies to brachytherapy sources as well as calibration and reference sources. The requirement to provide this confirmation is unreasonable and unnecessary and should be removed. In many cases, the primary container may be transferred from packaging provided by the manufacturer into another package that the pharmacy uses to transport orders to customers. For these cases, it is not possible for the applicant to confirm that *packaging will* not be altered. In situations where transfer of sources for redistribution from manufacturer's packaging involves the use of different shielding in pharmacy packaging, confirmation that *shielding will* not be altered is unnecessary and duplicative, especially in light of NRC's requirement under 10 CFR 32.72 and 8.10.12 of the guide to specify the shielding that is used.

NRC Staff Response: This comment has already been addressed in the response to the comment from Mallinckrodt, Inc., which is similar.

Location	Subject	Comment
Section 8.8.1	Dose Terminology	Section 8.8.1 starting on page 8-23 provides criteria and discussion regarding control of dose. Throughout this section the term "dose" is used in the context of the annual limit of 1 mSv. For the purpose of technical accuracy and to be consistent with the regulations that are referenced in this section, the term "dose" should be replaced with "total effective dose equivalent."

NRC Staff Response: Section 10 CFR 20.1003 defines "dose" as an acceptable generic term for "total effective dose equivalent." The term, as used, is accurate and consistent with the regulations pertaining to occupational exposure.

Location	Subject	Comment
Section 8.9	Description of Ventilation System	The response required under 8.9 on page 8-29 includes descriptions of ventilation and containment systems with airflow rates and pressure differentials. Also required is detail in diagrams that indicates specific locations of shielding, sources and other items "related to radiation safety."

While this draft, overall, takes a more risk-informed, performance-based approach to licensing nuclear pharmacies than in previous documents, section 8.9 requires information that is not performance-based and is unnecessarily restrictive in terms of the level of detail that would be included in the application. It is likely, with the detail provided, that subtle changes would later occur without any impact on the ability to meet NRC performance expectations and would require an amendment to the license. 8.9 should simply require the applicant to confirm that facilities and equipment will be sufficient to meet performance expectations such as 10 CFR 20.1301 and other relevant regulations. |

NRC Staff Response: This comment has already been addressed in the response to the comment from Mallinckrodt, Inc., which is similar.

Location	Subject	Comment
Section 8.10.1	Licensee Self-identified Deficiencies	The discussion in 8.10.1 on page 8-30 concerning licensee audits states that NRC will review the licensee's audit results and determine if corrective actions are satisfactory. The discussion goes on to state that audit findings would be included in the audit records that NRC will review. For a variety of reasons not related directly to NRC involvement, licensees may be unwilling to document specific observations made in the course of audits. If observations are recorded, some licensees may not be willing to make these available to outside agencies. NRC's insistence that this information be made available could ultimately be counterproductive in that audit reports may be documented in the format of a summary rather than a detailed roadmap of licensee performance. At the same time, if it is NRC's intent as reported to streamline and make the licensing and inspection programs more risk-informed and performance-based, then the focus of NRC's attention should be the end product or performance of licensee programs rather than licensees' internal audit reports and the information contained therein. The effectiveness of audit programs will be determined by NRC inspections, not by the information regarding observations in the audits themselves. If NRC's goal is to have licensees take prompt and corrective action to self-identified deficiencies, this can be achieved by licensee corrective action programs. The effectiveness of these programs and the documentation of their activities can be monitored without a review of audit reports. Regardless, it is the overall licensee performance that should ultimately be of interest to NRC and not the details of observations made in the course of internal audits.

NRC Staff Response: This comment has already been addressed in the response to the comment from Mallinckrodt, Inc., which is similar.

Location	Subject	Comment
Section 8.10.3, Table 8.1	Type B Packaging	Table 8.1 on page 8-34 contains labeling and survey requirements for packages of radioactive material. This table includes requirements for packages containing quantities greater than Type A. Since commercial nuclear pharmacies do not receive Type B shipments, this information is superfluous and should be removed from this table.

NRC Staff Response: This comment has already been addressed in the response to the comment from Mallinckrodt, Inc., which is similar.

Location	Subject	Comment
Section 8.10.5	Constraints on Air Emissions	The discussion in 8.10.5 on page 8-41 includes the following statements: "In addition, the licensee must control air emissions, such that the individual member of the public likely to receive the highest exposure does not exceed 0.1 mSv (10 mrem)(TEDE) per year from those emissions. If the exposure to a member of the public has exceeded the constraint on emissions, the licensee must report this exceedance, in accordance with 10 CFR 20.2203, and take prompt actions to ensure against recurrence." The intent of these statements is recognized yet the wording is confusing and technically incorrect. We recommend that these statements be reworded as follows: "In addition, the licensee must control air emissions, such that the individual member of the public likely to receive the highest total effective dose equivalent (TEDE) does not exceed the constraint level of 0.1 mSv (10 mrem) per year from those emissions. If exceeded, the licensee must report this in accordance with 10 CFR 20.2203, and take prompt actions to ensure against recurrence."

NRC Staff Response: We agree with the commenter's proposed wording. The text will be changed.

Location	Subject	Comment
Section 8.10.8	Dose Calibrator Beta-Correction Factors	The discussion and required response concerning assay of beta-emitting radionuclides in 8.10.8 makes no mention of the provision in 10 CFR 35.53(b) which exempts unit doses prepared by manufacturers and/or in accordance with 10 CFR 32.72 from the requirement for direct measurement of activity. This should be added to the discussion and provided as a response option in 8.10.8 and as a condition to the sample license provided as Appendix E.

NRC Staff Response: This comment has already been addressed in the response to the comment from Mallinckrodt, Inc., which is similar.

Location	Subject	Comment
Section 8.10.9	Type A Package Testing	On page 8-54 under 8.10.9 it states in the discussion that Type A packages must meet stringent criteria, including testing to "withstand accident situations and rough handling conditions." While we agree that these packages are robust enough to withstand rough handling and the most extreme conditions normally encountered in the course of commercial distribution, Type B containers and not Type A packages are designed to withstand accidents. Type B quantities of radioactive material are not handled within the scope of commercial nuclear pharmacy operations.

NRC Staff Response: We agree that Type B packages are specifically defined by DOT and NRC to withstand hypothetical accident conditions and that there is no mention of ability to withstand accidents in the DOT definition of Type A packages. When the draft guidance was written, we envisioned minor accidents in our description of the "robustness" of Type A packages, not the catastrophic accidents hypothesized for testing of Type B packages. The wording in Section 8.10.9 will be clarified to address the intended meaning.

Location	Subject	Comment
Section 8.10.12	Drug Versus Radiochemicals	8.10.12 starting on page 8-56 is titled "Radioactive Drug Shielding for Distribution." The use of the word "drug" in the title, discussion and applicant response section, as well as on page C-14 in Appendix C, requires information that is not performance-based and is unnecessarily restrictive in terms of the level of detail that would be included in the application implies that radiochemicals, sealed sources, radiobiologics and radiopharmaceuticals may be excluded. If it is NRC's intent to ensure all these materials are shielded adequately, then the title and wording should be revised to reflect this.

NRC Staff Response: Under 10 CFR 32.72(a)(3) applicants are required to provide information on the shielding provided by the packaging to show it is appropriate for the safe handling and storage of the radioactive drugs by medical use licensees. As used in this guidance document, the term "radioactive drug" is used interchangeably with the terms "radiopharmaceutical" and "radiobiologic"; therefore, no changes are needed.

Location	Subject	Comment
Section 8.10.12	Transport Radiation Shields	The required response from applicants on page 8-57 is to include the maximum content and the type and thickness of the "transportation radiation shield" for each type of container. This is another example of where the information is not performance-based and is unnecessarily restrictive in terms of the level of detail that would be included in the application. If dose rates or shielding need to be provided to ensure that dose rates are acceptable, then the applicant should be able to provide ranges or maximums that would not be exceeded to allow licensees to make modifications, including improvements, to packaging and shielding without having to amend a license to do so.

NRC Staff Response: This comment has already been addressed in the response to the comment from Mallinckrodt, Inc., which is similar.

Location	Subject	Comment
Section 8.10.13	Leak Testing of Sealed Sources	The discussion and response requirements included under 8.10.13 on page 8-57 and the conditions stated in the example license provided as Appendix E include leak test performance criteria but do not include the provision in 10 CFR 35.59(f)(1-5) which specifies sources for which leak testing is not required. This needs to be added to avoid the implication that all sealed sources must be leak tested.

NRC Staff Response: The provisions in 10 CFR 35.59(f) are not applicable to radiopharmacy licensees. The leak testing criteria for specific sealed sources is established in the Sealed Source and Device Registry certificate. If the conditions of the issuance of the registry certificate do not require leak testing, then the source is exempted from leak testing.

Location	Subject	Comment
Section 9	License Amendments	Section 9 on page 9-1 states that the licensee must submit an application for a license amendment before a change takes place if any of the information provided in the original application is modified or changed. While the need to keep licenses current is understood and appreciated, the requirement to amend a license for any changes is an unnecessary and unrealistic performance expectation placed upon both NRC staff and the licensee. This demand contravenes NRC initiative to streamline the licensing process and does not, at face value, provide licensees with the reasonable flexibility to make changes without amendment that do not have health and safety implications. A more reasonable approach would be the ability of the licensee to file reports that could be added to the licensee file to inform NRC of changes without the need to go through a formal amendment application and review process. Another option would be for NRC to establish a schedule of items typically specified in license applications that would or would not be subject to the amendment process if changes were made.

NRC Staff Response: Due to the nature of the new application format implemented for this NUREG, submission of amendment requests to address changes in licensee programs will likely become rare occurrences. Changes that will necessitate submission of an amendment request include, but are not limited to: modifications or additions to facilities; appointing a different individual to the position of Radiation Safety Officer; and adding new activities not previously authorized by the license (such as service activities). Changes to specific procedures will not require a license amendment prior to implementation due to the flexibility designed in the responses from the applicant.

Table U.5   Comment by the American Pharmaceutical Association, Dated June 29, 1999

Location	Subject	Comment
Numerous, and Specifically Sections 8.6.1, 8.8.3, and 8.10.8	Interference in the Practice of Pharmacy	To paraphrase, the commenter is concerned that the draft guidance document included suggested responses from applicants that would interfere in the practice of pharmacy, and lists Sections 8.6.1 (Distribution and Redistribution of Sealed and Unsealed materials); 8.8.3 (Instruction for Supervised Individuals Preparing radiopharmaceuticals); and 8.10.8 (Dosage Measurement Systems) as evidence of this interference.

NRC Staff Response: The writing team took special care to ensure that the draft guidance did not include anything that would interfere in the practice of pharmacy. Nowhere in the draft guidance is there discussion of specific pharmacy practices. The guidance document is intended to be used as an "implementation instrument," to aid applicants and licensees in meeting specific regulatory requirements, but it does not impose any new requirements that are not found in the regulations. Since the practice of pharmacy is outside NRC's regulatory jurisdiction, the guidance document is limited to the discussion of radiation safety practices to meet those regulatory requirements.

With regard to the sections referenced, the comment did not address specific aspects of the suggested responses from applicants that interfered with the practice of pharmacy. Furthermore, Section 8.8.3 (Instruction for Supervised Individuals Preparing Radiopharmaceuticals) does not require a response. Without more specific information, we cannot adequately address this comment.

Location	Subject	Comment
Entire Document	Coordination with Issuance of Proposed Revision to 10 CFR Part 35	It is our understanding that notable changes are currently taking place in 10 CFR 35, with significant changes to requirements for medical licensees. It is inappropriate to publish this NUREG document when it is not in line with the requirements of the upcoming 10 CFR 35 final rule. Although the final language and requirements of the new 10 CFR 35 are not yet known by licensees and members of the general public, enough is known about the intent of such changes to see that the suggested procedures in NUREG-1556 Vol. 13 are not aligned with those changes. It would be inappropriate for the procedures for commercial nuclear pharmacy licensees under Part 32 to be different from that required of medical licensees under Part 35. The two documents need to be coordinated in their requirements and/or suggested procedures for licensing documents.

NRC Staff Response: The only section in the proposed revision to 10 CFR Part 35 that will apply to radiopharmacies is the section on the definition and training requirements for an Authorized Nuclear Pharmacist, as is found in the current version of the regulation. Section 10 CFR Part 35 is concerned with the medical use of byproduct materials (i.e., the application of radiation and administration of radioactive materials to humans in the practice of medicine). Section 10 CFR 32.72, the "Radiopharmacy Rule," governs the manufacture and distribution of byproduct materials for medical use. Since the requirements of the current and proposed revision of 10 CFR Part 35, with the exception noted above, do not and will not apply to radiopharmacy licensees, it is not necessary to coordinate publication of the proposed guidance document in final form and the issuance of the proposed revision of 10 CFR Part 35.

APPENDIX V

Summary of Comments Received on Draft Revision 1 of NUREG-1556, Volume 13

Summary of Comments Received on Draft Revision 1 of NUREG-1556, Volume 13

For the tables in this Appendix, note that the page number reference associated with each comment under the location heading refers to the page number in the May 2007 NUREG-1556 Draft Report for Comment version of Volume 13, Rev.1, "Program-Specific Guidance About Commercial Radiopharmacy Licenses." Note that comments were requested on the specific changes in this NUREG related to the expanded definition of byproduct material and the NARM rule. Therefore, generally, only comments related to the NARM rule were considered. Comments that were related to other issues will be evaluated during any future revision of this NUREG.

Table V.1   Comment from Daniel J. Strom, Dated June 29, 2007

Location	Subject	Comment
Website	Bookmarks and hyperlinks on website	Rev. 0 is extremely useful in teaching nuclear pharmacy students, but only after bookmarks were added. Please implement bookmarks in Adobe Acrobat and create hyperlinks for contents, list of tables, etc.
NRC Staff Response: The NUREG-1556 series of documents are posted on the NRC website in PDF format, which allows search and thumbnail images of pages. Volume 13, Rev. 1, has been posted with a Table of Contents that will provide hyperlinks to each section of the NUREG. This comment will be considered in future revisions and posting of documents in the NUREG-1556 series.		

Table V.2   Comment from Michigan Department of Environmental Quality, Dated July 20, 2007

Location	Subject	Comment
General Comment	Public Dose	Radiochemical synthesis units using positron emission tomography (PET) radiopharmaceuticals release radioactive material to the air during their normal processes. The integrity of the transfer line or other hardware can catastrophically fail, releasing a bolus to the atmosphere. We strongly urge the Nuclear Regulatory Commission (NRC) to require PET radiopharmacies to submit an assessment of the potential doses to members of the public during routine use and during a catastrophic failure. We do not believe that the average NRC or state agreement inspector can adequately evaluate the ventilation system design and the computer modeling of public doses during a routine inspection. The complexity of the ventilation systems, the inherent limitations of the different computer codes, and the breadth of input data for the computer codes would be difficult for an inspector to evaluate during an on-site inspection. With the dose assessment submitted during licensing of the facility, NRC staff can adequately evaluate the premises and conclusions of the dose assessment. Then the inspector knows before the inspection that an annual release to the atmosphere of "x" curies of a radionuclide means a dose of "y" millirems to a member of the public. The inspector would need to verify during the inspection that the other input parameters in the dose assessment had not changed.

NRC Staff Response: The NRC staff must have sufficient information to make the necessary determination that the application meets the requirements in 10 CFR 30.33(a)(2) which in this case means the ventilation system will be adequate for the licensee to meet the requirements in 10 CFR 20.1101. The NRC evaluation will include both normal and equipment failure conditions. The NRC does not provide prescriptive guidance because of the flexibility the applicant has in designing facilities, choosing ventilation systems, and developing procedures to meet the requirements. It is the applicant's responsibility to provide sufficient information. During an NRC inspection, the licensee must be able to demonstrate, by measurement or calculation, that the annual dose limits for members of the public has not been exceeded. No change was made to the guidance document.

Location	Subject	Comment
General Comment	Referenced ANSI Standards	The ANSI standards referenced in Regulatory Guide 8.37, "ALARA Levels for Effluents from Materials Facilities" and Regulatory Guide 4.20, "Constraints on Release of Airborne Radioactive Materials to the Environment for Licensees Other Than Power Reactors," have been revised. These Regulatory Guides should be reviewed and revised. • ANSI N42.18 "Specification and Performance of On-Site Instrumentation for Continuously Monitoring Radioactivity in Effluents" was revised in 2004. • ANSI N13.1 "Guide to Sampling Airborne Radioactive Materials in Nuclear Facilities" was revised in 1999 and renamed "Sampling and Monitoring Releases of Airborne Radioactive Substances From the Stacks and Ducts of Nuclear Facilities."

NRC Staff Response: The revised ANSI standards N42.18 and N13.1 have been reviewed and the reference in Appendix M of this guidance document has been updated to reflect the revised standards. Revision to Regulatory Guides 8.37 and 4.20 is beyond the scope of this guidance document revision.

Location	Subject	Comment
Abbreviations (Page xv)	Abbreviations	Abbreviations - Add the following: DU Depleted Uranium LSC Liquid Scintillation Counter NaI Sodium Iodide NaI (Tl) Sodium Iodide (thallium activated) rad Unit of Absorbed Dose gy Gray-SI unit of absorbed dose rem roentgen equivalent man And delete: cm centimeter mGy milliGray mR milliroentgen mrem millirem mrem/hr millirem per hour mSv millisievert mSv/hr millisievert per hour And add SI prefixes: Prefix Symbol Factor Examples micro μ 10-6 μR Milli m 10-3 mCi, mR Centi c 10-2 cm Kilo k 10+3 kg, kBq mega M 10+6 MBq Giga G 10+9 GBq Tera T 10+12 TBq

NRC Staff Response: The current Abbreviations section is consistent with NRC policy and the guidance documents in the NUREG-1556 series. Therefore, not all of the suggested changes have been made.

Location	Subject	Comment
Section 8.9 & Appendix C (Pages 8-31 and C-9)	Facilities and Equipment	"Response from Applicant" regarding "Facilities and Equipment." This section states, "Verification that ventilation systems ensure that effluents are ALARA, are within the dose limits of 10 CFR 20.1301, and are within the ALARA constraints for air emissions established under 10 CFR 20.1101 (d)." What would be considered sufficient verification? Does a facility need to submit a computer model calculating the projected doses to members of the public at various nearby locations or will an unsupported statement that public doses are ALARA be considered sufficient?

NRC Staff Response: The applicant must provide sufficient information for NRC staff to determine that the application meets the requirements in 10 CFR 30.33. This information could include computer model calculations or measurements to verify that effluents from the facility will be ALARA. A single statement that public doses are ALARA would not be sufficient. Note that the table in Appendix C is a checklist that duplicates the response to text found in the main body (Chapter 8) of this document.

Location	Subject	Comment
Appendix J	Radiation Monitoring	"Radiation Monitoring Instrument Specifications and Model Survey Instrument Calibration Program" should include a discussion on the calibration of radiation detection equipment installed to monitor and quantify the activity released to the atmosphere. For PET radiopharmacies, stack exhaust monitors may be sodium iodide detectors mounted adjacent to the exhaust system. They are calibrated by releasing a known millicurie quantity of radioactive material. The number of counts above background can then be correlated with a known activity. This guidance document should state if the NRC will require subsequent periodic releases to annually (quarterly, monthly) "calibrate" these monitors or will the NRC accept a procedure using check sources to confirm that the response to the check sources has not changed since the initial calibration.

NRC Staff Response: In 10 CFR Parts 20, 30, and 32, there are no specific requirements for how and when radiation monitoring instruments are calibrated. Instruments should be calibrated in accordance with the instrument manufacturer's recommendations. Therefore, specific guidance on the calibration of air monitoring instruments is not provided in this document.

Location	Subject	Comment
Appendix K	Public Dose	"Public Dose" should mention that air intakes for the radiopharmacy building and for adjacent buildings need to be considered in the evaluation of doses to members of the public due to atmospheric releases.

NRC Staff Response: Appendix K currently provides general guidance on the methods that could be used for determining radiation doses to members of the general public. The NRC staff believes that the information in this Appendix is adequate. Therefore, no additional specific information needs to be added to this Appendix.

Location	Subject	Comment
Appendix R (Page R-4)	Airborne Effluent Release Monitoring	"Air Stack Release Monitoring." ANSI N13.1 "Guide to Sampling Airborne Radioactive Materials in Nuclear Facilities" was revised in 1999 and renamed "Sampling and Monitoring Releases of Airborne Radioactive Substances From the Stacks and Ducts of Nuclear Facilities."

NRC Staff Response: The revised ANSI Standard N13.1 has been reviewed and the reference in Appendix R of this guidance document has been updated to reflect the revised standards.

Location	Subject	Comment
Appendix R (Page R-6)	Airborne Effluent Release Monitoring	"References." • ANSI N13.1 "Guide to Sampling Airborne Radioactive Materials in Nuclear Facilities" was revised in 1999 and renamed "Sampling and Monitoring Releases of Airborne Radioactive Substances From the Stacks and Ducts of Nuclear Facilities." • ANSI N42.18 "Specification and Performance of On-Site Instrumentation for Continuously Monitoring Radioactivity in Effluents" was revised in 2004.

NRC Staff Response: The revised ANSI Standards, N13.1 and N42.18, have been reviewed and the references in Appendix R of this guidance document have been updated to reflect the revised standards.

Location	Subject	Comment
Entire Document	General Comment	Spelling errors were noted.

NRC Staff Response: The noted spelling errors have been corrected.

Table V.3   Comments from Darrell R. Fisher, Dated July 30, 2007

Location	Subject	Comment
Section 8.6.1 (Page 8-11)	Radioactive Drugs	In the definition of radioactive drugs, radiobiologics [radio Latin emitting rays; bio, bios Greek life, living; logics, logica Latin of reason, guiding principles] seems to be an incorrect choice of word and incorrect usage in this context. Monoclonal antibodies are non-living chemicals that may be considered as biological agents or radioactive drugs because they function in a certain way in living systems by seeking out cell-surface antigens, but the term radiobiologics is not a standard synonym for radiolabeled monoclonal antibody.

NRC Staff Response: The term "biologics" was used appropriately to indicate a biological product (e.g., monoclonal antibodies, or Tc-99m tagged red blood cells). The sentence clarifies that the term "radioactive drugs" has a broader meaning than radiopharmaceuticals because it also includes the radiobiologics regulated by FDA under its biologic license application process. Therefore, no change was made to the guidance.

Location	Subject	Comment
Section 8.6.1 (Page 8-12)	Redistribution of Discrete Sources of Ra-226	If discrete Ra-226 sources are to be redistributed for beneficial reuse and reconfigured as targets for accelerator irradiation to produce new radioactive materials, the requirements in this section appear to be overly restrictive and inhibitive of this practice. For redistribution of discrete sources of radium-226, it may be impossible to confirm that the discrete sources of radium-226 will be obtained by a [or from a??] manufacturer authorized to distribute it. For most legacy sources, it will not be possible to identify the manufacturer. Manufacturer-supplied package inserts may not have been produced. Limitations on the ability of a licensee to alter Ra-226 packaging may prevent Ra-226 from being recombined into larger-activity sources for use in configurations that are necessary to use Ra-226 as a target for new isotope production. An example would be the use of Ra-226 to produce Ra-225, which decays to Bi-213 for medical applications.

NRC Staff Response: The redistribution of discrete sources of radium-226 (Ra-226) in this section is not referring to reconfiguring the Ra-226 as targets for accelerator irradiation. The redistribution of discrete sources of Ra-226 in this guidance document refers to discrete sources intended to be distributed under 10 CFR 32.74 to a medical use licensee that are distributed to a commercial radiopharmacy, which in turn, distributes them to a medical use licensee. The Ra-226 discrete source is then used by the medical use licensee for calibration of an instrument(s) or for medical use.

Location	Subject	Comment
Section 8.7 (Page 8-16)	Management	The commentary that "management responsibility and liability are sometimes underemphasized or not addressed in applications and are often poorly understood by licensee employees and managers" appears to be inappropriate and unnecessary in this guidance document. It should not be the purpose of this document to assume the competence of some applicant organizations. Delete text.

NRC Staff Response: This comment is not related to the NARM rule and, therefore, is beyond the scope of this guidance document revision. This comment will be evaluated during any future revision of this NUREG.

Location	Subject	Comment
Section 8.7.3 (Page 8-22)	Authorized Users	The statement that "applicants should pay particular attention to the type of radiation involved...For example, someone experienced with gamma emitters may not have appropriate experience for high-energy beta emitters" seems unnecessary if the student has met the requirements in the text above and has studied the characteristics of ionizing radiation. Again, the NRC appears to be judging competency based on the assumption that a situation could exist where a trained authorized user understands gamma rays but not beta particles. Delete text.

NRC Staff Response: This comment is not related to the NARM rule and, therefore is beyond the scope of this guidance document revision. This comment will be evaluated during any future revision of this NUREG.

Location	Subject	Comment
Section 8.10.6 (Page 8-45)	Safe Use of Radionuclides	The Guidance assumes the radiopharmacy uses only Mo-99/Tc-99m generator systems, when many other types of generators are available or could be developed in the future. The elution breakthrough test is applicable to any radionuclide generator system in the radiopharmacy, not just Mo/Tc. Examples of other generator systems include: Sr-82/Ru-82, Sr-90/Y-90, Ac-225/Bi-215, Ac-227/Ra-223, and Ge-68/Ga-68.

NRC Staff Response: The NRC only has promulgated specific breakthrough test requirements for molybdenum-99/technetium-99m and strontium-82/rubidium-82 generator systems under 10 CFR 30.34(g). However, the strontium-82/rubidium-82 generator breakthrough test is not generally performed at the pharmacy, but at the medical facility prior to first patient use. Therefore, this guidance document only refers to the molybdenum-99 breakthrough measurements.

Table V.4 Comments from CORAR, Dated August 1, 2007

Location	Subject	Comment
General Comment	Authorized Nuclear Pharmacist	One general comment concerns the issue of "grandfathering" of authorized nuclear pharmacists who, as discussed in the proposed rule published on July 28, 2006, "will not be required to meet new training and experience requirements as long as their duties and responsibilities under the new license do not significantly change." There is no such discussion of grandfathering in the draft Volume 13, and this omission could be critical to ensuring the continued supply of accelerator produced radiopharmaceuticals. Specific guidance on grandfathering of authorized nuclear pharmacists should be added to section 8.7.2 and any other relevant sections of the draft NUREG.

NRC Staff Response: Guidance regarding "grandfathering" of nuclear pharmacists has been added in Section 8.7.2.

Location	Subject	Comment
Section 8.2 (Page 8-2)	Timely Notification of Transfer of Control	It is often difficult or impossible for licensees to meet this requirement, as often the RSO is not at a level to be made aware of such a change in the business prior to its execution. The best that can be expected in some cases is for immediate notification to be made when the RSO is made aware of the change.

NRC Staff Response: As discussed in this section, it is the licensee's responsibility, not the Radiation Safety Officer's responsibility, to provide written notification prior to transferring control of the license. No change was made to this section.

Location	Subject	Comment
Section 8.5.2 (Page 8-9)	Financial Assurance	CORAR agrees with the statement in this section that "most radiopharmacy applicants and licensees do not need to take any action to comply with the financial assurance requirements" because they possess radionuclides that have half-lives no greater than 120 days. We believe that it would be very useful to licensees for NRC to add to this discussion a statement to the effect that a decommission plan would also not be needed. This also may be particularly relevant in some states where licensees had been required to establish a plan, regardless of the need to obtain financial surety, for activities such as pharmacy renovation involving only a portion of the facility. NRC decommissioning requirements need to be a matter of strict compatibility for Agreement States.

NRC Staff Response: The NRC does have requirements for submitting a decommissioning plan in 10 CFR 30.36 (g)(1) that may apply to radiopharmacy applicants. Therefore, it would not be accurate to indicate that a decommissioning plan would not be required. Also, changes to NRC's decommissioning compatibility requirements for Agreement States are beyond the scope of this guidance document revision. Therefore, no change was made to the guidance document.

Location	Subject	Comment
Sections 8.6.1 and 8.6.2 (Pages 8-11 and 8-14)	Preparation and Distribution of Radioactive Drugs	CORAR commented in 1999 on the original draft of Vol. 13 that the discussion in this section needed to include the characterization of the compounding of non-FDA approved radiochemicals as a nuclear pharmacy, and that NRC should state a position on acceptability of this practice. NRC responded in Appendix U of the proposed draft Vol. 13 that "fitness of a particular radiochemical for use in compounding radiopharmaceuticals for ultimate use in medicine is outside NRC's regulatory authority, and therefore, discussion of this issue is not appropriate in this guidance document."

CORAR maintains that it is within the scope of NRC's regulatory authority to require a license to manufacture and distribute radiopharmaceuticals where an operation is using non-FDA approved radiochemicals to compound "radiopharmaceuticals" and FDA considers this subject to "manufacturing" requirements rather than within the scope of pharmacy practice. |

NRC Staff Response: The NRC clearly states in 10 CFR 32.72(d) that nothing in 10 CFR 32.72 relieves the licensee from complying with applicable FDA, other Federal, and State requirements governing radioactive drugs. The NRC considers the fitness of a particular radiochemical (whether FDA-approved or not) within the regulatory purview of the FDA and State boards of pharmacy to resolve and is beyond the scope of NRC's regulatory authority.

Location	Subject	Comment
Section 8.6.1 (Page 8-11)	Distribution of Radioactive Material	CORAR recommends that discussion be added to this section to address the transfer of radioactive material from nuclear pharmacies to mobile nuclear medicine operations at temporary locations other than those specifically listed on a radioactive material license.

NRC Staff Response: This comment is not related to the NARM rule and, therefore, is beyond the scope of this guidance document revision. This comment will be evaluated during any future revision of this NUREG.

Location	Subject	Comment
Section 8.6.1 (Page 8-13)	Redistribution of Sealed Sources	CORAR commented in 1999 on the original draft of Vol. 13 that it opposed the requirement for an applicant to confirm that the manufacturer's labeling and packaging will not be altered for redistribution of sealed sources, as an unnecessary burden on nuclear pharmacies. NRC responded in Appendix U of the proposed draft Vol.13 with the statement that "if the packaging is not specified in the approval for initial distribution, then other persons may repackage the source or device for redistribution." CORAR suggests that NRC add this statement to section 8.6.1 of NUREG-1556, Vol. 13.

NRC Staff Response: This comment is not related to the NARM rule and, therefore, is beyond the scope of this guidance document revision. This comment will be evaluated during any future revision of this NUREG.

Location	Subject	Comment
Section 8.9.2 (Page 8-30)	Facilities and Equipment for PET Radiopharmacies	In the discussion it states, "the majority of the radioactive effluents at a PET radiopharmacy are produced during the synthesis of the PET radiopharmaceutical." A reference is also made in this section to Appendix R as it provides more information on effluent monitoring. CORAR agrees that at least in some cases PET radionuclides do contribute to the profile of radioactive gaseous effluents. However, with the highlighted discussion here on PET, NRC should provide some detailed guidance on monitoring PET effluents and demonstrating compliance with relevant limits. There also is no such guidance in Appendix R.

NRC Staff Response: The NRC staff must have sufficient information to make the necessary determination that the application meets the requirements in 10 CFR 30.33(a) which in this case means that the equipment and procedures used to monitor effluent releases meet the requirements in 10 CFR Part 20. The NRC does not provide prescriptive guidance on monitoring for PET or other radionuclides because of the flexibility the applicant has in facility design, effluent monitoring equipment, and procedures. It is the applicant's responsibility to provide sufficient information. No change was made to the guidance.

Location	Subject	Comment
Section 8.10.1 (Page 8-31)	Audit Program	CORAR members have operations that are subject to the regulatory requirement to conduct annual audits of their Radiation Protection Programs. CORAR commented in 1999 on the original draft of Vol. 13 that it is imperative that NRC recognizes the efforts of a licensee to identify and take appropriate actions for self identified deficiencies and not to penalize the licensee for its pro-active regulatory compliance program. NRC responded in Appendix U of NUREG-1556, Vol. 13, by stating that NRC enforcement policy (NUREG-1600) specifically affords inspectors the authority to withhold the issuance of a Notice of Violation for licensee identified violations in those cases where it is warranted and appropriated. CORAR appreciates this position but believes it does not go far enough because of the subjective nature of applicability and ongoing exposure of licensees to judgmental variability between inspectors. CORAR has addressed this issue separately with NRC in March 2007 in response to NRC Enforcement Policy; Proposed Plan for Major Revision, Federal Register volume 72, No. 16, page 3429, January 25, 2007. At that time CORAR commented that the Policy should address issues involving licensee disclosure of findings and other information as a result of audits conducted independent of NRC inspections. With regard to audits conducted by or on behalf of licensees, NRC should not require that the results of such audits be disclosed nor should NRC inspectors request copies of audit reports or findings. In addition, audit reports or findings should not be used by NRC to trigger NRC enforcement investigations.

NRC Staff Response: This comment is not related to the NARM rule and, therefore, is beyond the scope of this guidance document revision. This comment will be evaluated during any future revision of this NUREG.

Location	Subject	Comment
Section 8.10.1 (Page 8-31)	Audit Program	In addition to the relief from civil penalty provided for Severity Level I – III violations in the current Policy, NRC should not cite a Notice of Violation for any non-reportable compliance problems self-identified and promptly and effectively corrected by the licensee. It would be reasonable for NRC to expect the finding, identification of root cause, and corrective action to be documented by the licensee for future reference. Alternatively, NRC could disposition these as Non-Cited Violations. NRC should ensure that discussion in section 8.10.1 of NUREG-1556, Vol. 13 reflect these recommendations. Reference to the Enforcement Policy should be maintained so that any revisions to it will be incorporated by reference into this licensee guidance document.

NRC Staff Response: The recommendations regarding changes to this section are not related to the NARM rule and are beyond the scope of this guidance document revision. This comment will be evaluated during any future revision of this NUREG.

Location	Subject	Comment
Section 8.10.2 (Page 8-35)	Radiation Monitoring Instruments	NRC in this section suggests that an applicant may respond with a statement that equipment used will meet the radiation monitoring specification published in Appendix J. Table J-1 in Appendix J includes a list of instrument types and "specifications" intended to "help applicants and licensees choose the proper radiation detection equipment for monitoring the radiological conditions at their facilities." However, a review of Table J.1 concludes that there really aren't any useful specifications provided. For example, energy ranges specified are "all energies." Efficiencies are specified as "moderate" or "high." These are very general and non-specific terms. We recommend that NRC include a table that includes real specifications that would be more useful to those who need this level of technical guidance.

NRC Staff Response: The information in this section and Appendix J was never intended to provide specific information such as specific energy ranges. Therefore, the title of Appendix J and the text in Section 8.10.2 have been revised to better describe the general information available in the Appendix.

Location	Subject	Comment
Section 8.10.3 (Page 8-39)	Record Maintenance	Table 8.2 should be expanded to include the retention of written directives for three years in accordance with 35.2040-2041.

NRC Staff Response: Records for written directives are required to be maintained by the medical-use licensee in accordance with 10 CFR Part 35 and not a commercial radiopharmacy licensee. Therefore, the retention of written directives was not added to Table 8.2.

Location	Subject	Comment
Section 8.10.4 (Page 8-40)	Occupational Dose	NRC in recent years has paid significant attention to the issue of extremity dose and occupational monitoring at commercial nuclear pharmacies. CORAR and its members have approached NRC and have established a partnership in an effort to investigate the issue and develop needed guidance on methodologies for monitoring extremity dose to demonstrate compliance with 20.1201(a)(2)(ii). CORAR believes that guidance on extremity dose monitoring is warranted and strongly recommends that this section include discussion on this.

NRC Staff Response: The current guidance in this section provides general information regarding occupational dose requirements. Specific guidance on the methodologies for monitoring extremity dose is beyond the scope of this document revision as it does not relate to the NARM rule. This comment will be considered during any future revision of this NUREG.

Location	Subject	Comment
Section 8.10.4 (Page 8-41)	Occupational Dose	This section should provide some guidance on whether an evaluation conducted to determine that an individual's dose is not likely to exceed 10% of the applicable limit needs to be conducted initially or at a recurring (e.g., annual) frequency thereafter. CORAR believes that the evaluation only needs to be conducted initially unless there is a change in the procedure or operation that could result in a higher exposure.

NRC Staff Response: The current guidance in this section indicates that an evaluation of the dose an individual is likely to receive should be performed prior to allowing the individual to receive a dose and does not indicate that the evaluation should be performed at a recurring frequency thereafter. As indicated in Regulatory Guide 8.34, "Monitoring Criteria and Methods to Calculate Occupational Doses," which is referenced in this section, if an individual's radiation exposure conditions change, the need to provide individual monitoring should be reevaluated.

Location	Subject	Comment
Section 8.10.5 (Page 8-44)	Public Dose	There is discussion in this section on the need for licensees to control air emissions so that the constraint level of 0.1 mSv is not exceeded. However, there is no mention in this section of methods acceptable to NRC to demonstrate compliance with the constraint level. CORAR recommends that NRC provide in this section an acceptable method (e.g. EPA COMPLY code), or make reference to other NRC guidance that provides a method for demonstrating compliance with the constraint level.

NRC Staff Response: A reference to Regulatory Guide 4.20, "Constraints on Release of Airborne Radioactive Materials to the Environment for Licensees Other Than Power Reactors," was added to Appendix K. This regulatory guide provides guidance on acceptable methods that can be used to demonstrate compliance with the air emissions constraint level.

Location	Subject	Comment
Section 8.10.6 (Page 8-44)	Safe Use of Radionuclides	Discussion in this section states, "licensees are responsible for the security and safe use of all licensed material from the time it arrives." CORAR recommends that NRC clarify the distinction between delivery of radioactive material by the carrier and receipt by the authorized consignee. This has implications with respect to the security of material in transport and obligations to report lost or missing shipments of radioactive material. It would be helpful for NRC to specify, or provide a reference that specifies, when a transfer from one licensee to another has been completed and at what point is security of the material transferred from the consignor to the consignee. It has been clarified by U.S. DOT in 49 CFR 171.8 regarding the definition of "unloading incidental to movement" that the cycle of transportation ends when delivery is made. This needs to be taken into consideration by NRC for additional discussion in this section.

NRC Staff Response: This comment is not related to the NARM rule and, therefore, is beyond the scope of this guidance document revision. This comment will be evaluated during any future revision of this NUREG.

Location	Subject	Comment
Section 8.10.6 (Page 8-45, Figure 8-8)	Use of Appropriate Shielding	The picture intends to show the use of appropriate shielding in a nuclear pharmacy operation. Compared to actual nuclear pharmacy operations, it suggests a situation that does not employ best practices with regard to ALARA. For example, there are multiple unshielded containers in proximity to the extremities and no evidence of any remote or extended handling devices within reach. The handling is also done on a bench top that would be unacceptable for dispensing of radiopharmaceuticals. This picture should be left out of the guidance or replaced with a more acceptable example.

NRC Staff Response: This comment is not related to the NARM rule and, therefore, is beyond the scope of this guidance document revision. This comment will be evaluated during any future revision of this NUREG. Note that this figure does represent the appropriate shielding for using/dispensing some radiopharmaceuticals (e.g., technetium-99m).

Location	Subject	Comment
Entire Document	General Comment	The term "radionuclides" instead of "radioisotopes" should be used here and throughout the document.

NRC Staff Response: The term "radioisotope(s)" has been changed to "radionuclide" when applicable.

Location	Subject	Comment
Section 8.10.7 (Page 8-49, Figure 8.10)	Radiation Surveys	The figure shows improper monitoring technique. The detector needs to be placed as close to the object being surveyed without making contact.

NRC Staff Response: This figure is meant only to illustrate that generally, users of unsealed licensed material should survey themselves before leaving restricted areas. Therefore, no change was made to this figure.

Location	Subject	Comment
Section 8.11 (Page 8-60)	Disposal by Decay-in-Storage	NRC suggests that waste held for decay should be held until a date when "ten half-lives of the longest-lived radioisotope have transpired." Other recent NRC guidance has dropped this requirement and only requires that residual radioactivity be determined to be indistinguishable from background prior to disposal. The guidance in this section should be made consistent with other NRC guidance.

NRC Staff Response: This text has been removed from this section, as the NRC staff agrees that waste should be held for decay until the radiation exposure rate cannot be distinguished from background radiation levels.

Location	Subject	Comment
Appendix K (Page K-3)	Occupancy Factors	CORAR recommends that NRC incorporate into Table K.1 the occupancy factors from NCRP Report 147 (page 31) for planning and assessing public dose.

NRC Staff Response: This comment is not related to the NARM rule and, therefore, is beyond the scope of this guidance document revision. This comment will be evaluated during any future revision of this NUREG.

Location	Subject	Comment
Appendix R (Page R-6)	Air Stack Release Monitoring	The reference to ANSI N13.1 (1969) should be revised to refer to the updated 1999 version.

NRC Staff Response: The revised ANSI N13.1 has been reviewed and the reference in Appendix R of this guidance document has been updated to reflect the revised standards.

Table V.5   Comments from Washington State Department of Health, Dated August 1, 2007

Location	Subject	Comment
Foreword (Page x)	General	The second paragraph, 3rd sentence should read: This expanded definition includes the material that is **produced**, extracted or converted after extraction for use for a commercial, medical, or research activity.

NRC Staff Response: The word "produced" has been added to this text.

Location	Subject	Comment
Section 8.5.1 (Page 8-5)	Sealed Sources or Devices	The fourth paragraph reads: It should also be noted that NRC's regulatory authority includes the new byproduct material produced prior to August 8, 2005. As a result, neither NRC, an Agreement State, or a non-Agreement State may have performed a safety evaluation of the sealed source or device. Therefore, the sealed source or device may not have an Sealed Source and Device Registry (SSDR) registration certificate. 10 CFR 30.32(g) provides information that must be submitted for these types of sources. This paragraph is written poorly and the intent of the paragraph is unclear. The paragraph should be rewritten to clearly express the intent.

NRC Staff Response: This paragraph has been edited to clarify its intent.

Location	Subject	Comment
Section 8.9.2 (Page 8-30)	Shielding/ Remote Handling Equipment	The guidance should separate the shielding discussion from the remote handling equipment discussion. Descriptions of shielding should be provided for the transfer lines when transferring material from the cyclotron to the hot cell, and between the hot cell and the chemistry synthesis unit. Descriptions of shielding should also be provided for the physical hot cell, chemistry synthesis unit, both short lived and long lived (from target rebuilding) waste. The applicant should describe the remote handling equipment that will be used (i.e. manipulators in the hot cell, automatic transfer lines to move material between process stations).

NRC Staff Response: The discussion on shielding has been edited to differentiate between the description of the type of shielding and the remote handling equipment that will be used.

Location	Subject	Comment
Section 8.9.2 (Page 8-30)	Effluent Control and Monitoring	Examples of engineered controls to reduce the amount of material released should include the use of gas-trapping bags to capture the effluent from the chemistry synthesis unit. It is a common practice to hold up the high activity, short lived effluent in a bag for decay. This method is extremely effective in substantially reducing the amount of activity released as effluent to the air from a manufacturer of PET radiopharmaceuticals. If gas-trapping bags are used the applicant must address the location, shielding, and handling (emptying) of these gas-trapping bags.

NRC Staff Response: In the discussion section of Section 8.9.2, the use of a containment system for the decay of effluents was mentioned. Holding and decaying short-lived effluents may be done in many different ways. Therefore, specific details on the type of containment system used (e.g., gas-trapping bags) were not added to this discussion.

Location	Subject	Comment
Section 8.9.2 (Page 8-30)	Facilities and Equipment	The applicant should also discuss the procedures/ controls in place to assure the integrity of the transfer lines are not compromised prior to a transfer. A loss of material during a transfer from the cyclotron to the hot cell could result in a substantial amount of high activity material being spilled, potentially causing a high personnel exposure.

NRC Staff Response: The NRC does not require the applicant to provide specific procedures on operation/use of the equipment. However, the applicant should have, as part of its Radiation Safety Program, procedures that would prevent the loss of radioactive material.

Location	Subject	Comment
Section 8.9.2 (Page 8-32)	Facilities and Equipment	Figure 8.4 should appear in Section 8.9 not 8.10; Figure 8.4 should be moved to page 8-31 and the information on page 8-31 should be moved to page 8-32.

NRC Staff Response: Figure 8.4 has been moved into Section 8.9.2, "Facilities and Equipment for PET Radiopharmacies."

Location	Subject	Comment
Appendix M	Department of Transportation Requirements	Why was the information in Appendix M replaced completely? The Appendix M information contained in the original NUREG 1556 vol 13 appeared to be informative and beneficial.
NRC Staff Response: The previous information in Appendix M was revised to help ensure that the applicant would have current transportation information as provided in the Department of Transportation (DOT) regulations. The previous information in Appendix M had very specific information regarding DOT regulations that may change, and NRC would not be able to ensure that any changes to this information could be immediately incorporated into this Appendix.		

Table V.6 Comments From Washington University in St. Louis, Dated August 1, 2007

Location	Subject	Comment
Chapter 1 (Page 1-1)	Purpose of Report	In the first paragraph on page 1 - 1, the draft guidance states ", . .the phrases or terms, 'commercial radiopharmacy,' 'radiopharmacy,' 'nuclear pharmacy,' and 'pharmacy' are used interchangeably." We strongly recommend that NRC not include "pharmacy" as one of these interchangeable terms. We also recommend that a clarification statement be added noting that the interchangeable use of "commercial radiopharmacy," "radiopharmacy" and "nuclear pharmacy" does not necessarily mean the guidance applies for a non-commercial radiopharmacy or a non-commercial nuclear pharmacy.
NRC Staff Response: The term "pharmacy" was deleted from the list of interchangeable terms. As discussed in response to the next comment, the term "pharmacy" was replaced with the term "nuclear pharmacy" in the text where the change was appropriate. Also, the first sentence of this section indicates that this guidance document is for an applicant that is applying for a commercial radiopharmacy license. Therefore, additional text is not needed to indicate that this guidance is not for a noncommercial radiopharmacy.		

Location	Subject	Comment
Section 8.9	Facilities and Equipment	We recommend replacing all uses of the term "pharmacy" with "nuclear pharmacy," "radiopharmacy," or "pharmacy (radiopharmaceuticals)." The following examples show where the use of the term, "pharmacy", is giving either incorrect or unclear guidance. Without use of a clarifying term such as "nuclear pharmacy," "radiopharmacy," or "pharmacy (radiopharmaceuticals)," the following statements imply a state pharmacy license is appropriate to become a commercial radiopharmacy: **Page 8-26** – "Licensure as a pharmacy by a State Board of Pharmacy; or…" **Page 8-27** – "Applicants must provide: Copies of their registration or license from a State Board of Pharmacy as a pharmacy…" **Page 8-30** – "PET radiopharmacies must demonstrate that they are …licensed as a pharmacy by the State's Board of Pharmacy…" **Page C-9** – "Provide a copy of the registration or license from a State Board of Pharmacy as a pharmacy…" Without the use of "nuclear pharmacy" or "radiopharmacy," the following statement may be confusing by suggesting an individual only needs pharmacy experience: **Pages 8-20, C-7 & D-5** – "The individual practiced at a pharmacy at a Government agency or Federally recognized Indian tribe before April 8, 2007, or at all other pharmacies before August 8, 2009, or an earlier date as noticed by NRC."

NRC Staff Response: The term "pharmacy" was replaced with the term "nuclear pharmacy" where applicable. It should be noted that most State Boards of Pharmacy only issue pharmacy licenses and not "nuclear" pharmacy licenses. Therefore, all uses of the term "pharmacy" were not replaced in the document.

Location	Subject	Comment
Section 8.7.2 (Page 8-19)	Authorized Nuclear Pharmacist	In Section 8.7.2 Discussion, the draft guidance describes the sections of regulation defining the training and experience requirements to become an Authorized Nuclear Pharmacist (ANP) at a commercial radiopharmacy. We recommend that a statement be added to this section which discusses the "grandfathering" of a nuclear pharmacist who has used only accelerator-produced radioactive materials, discrete sources of Ra-226, or both for medical or nuclear pharmacy uses. We suggest the following paragraph be added at the end of this discussion section: "Nuclear pharmacists who used accelerator-produced radionuclides or discrete sources of Ra-226 during the effective period of the waiver do not have to meet the requirements of 10 CFR 35.59, or the training and experience requirements in 10 CFR Part 35, Subpart B for those materials and uses. The criteria for such nuclear pharmacists are described in 10 CFR 32.72(b)(4) and acceptable documentation is discussed in Appendix G."

NRC Staff Response: Text has been added to this section to discuss the "grandfathering" of a nuclear pharmacist who has used only accelerator-produced radioactive materials for the preparation of radioactive drugs.

Location	Subject	Comment
Appendix G (Page G-5)	Training Documentation	The statement under "State or Territory where Licensed" on page G-5 indicates that pharmacists are licensed to prescribe drugs. This statement is incorrect, and we recommend the statement be corrected to say that pharmacists are licensed to dispense drugs.

NRC Staff Response: The statement has been corrected to say that pharmacists are licensed to practice pharmacy.

Location	Subject	Comment
Appendix G (Page G-6)	Training & Experience Documentation	In Appendix G Part II. Preceptor Attestation (page G-6), the current regulatory definition of preceptor is quoted, and we note that nowhere is it indicated that the preceptor must have the same "authorization" as is sought by the individual whose training and experience is being verified by the preceptor. As NRC is preparing to "grandfather" individuals who have used accelerator-produced radionuclides to be an ANP (or an AU, AMP or RSO), there is an opportunity to bring the training and experience criteria for ANPs (AUs, AMPs and RSOs) more in line with the preceptor definition. We agree that a preceptor statement from a current ANP is appropriate for those individuals seeking to become an ANP by the alternative pathway. WU strongly recommends that the NRC Staff and, in particular, the Nuclear Regulatory Commissioners reconsider the need for an ANP preceptor statement for those individuals who are board-certified by an NRC-recognized specialty board. Each of the specialty boards recognized by the NRC have proven to the NRC that their board eligible candidates meet the training and experience requirements for the type(s) of medical use for which they are recognized. In order to sit for a board exam, an individual requires the recommendation of a sponsor who verifies the individual has met all of the requirements to become board-certified. While this sponsor may not be an ANP, the sponsor is responsible to the board for recommending only individuals who meet the board's, and therefore the NRC's, requirements. Successful completion of the board exam by the individual gives further verification of the individual's training and experience. WU believes the current regulations imposing the additional requirement of an ANP preceptor statement is an unnecessary redundancy that has greatly complicated the process of approving an individual as an ANP, and has led to the trivialization of long-established radiopharmacy board-certification.

NRC Staff Response: Any revisions to the training and experience requirements would require a revision to NRC's current regulations. Therefore, this comment is beyond the scope of this guidance document revision.

Location	Subject	Comment
Appendix G (Page G-6)	Training & Experience Documentation	We appreciate that NRC has taken care to ensure the continuing access of PET imaging techniques by allowing the "grandfathering" of individuals who have used accelerator-produced radionuclides to become ANPs (or AUs, AMPs or RSOs). We believe that NRC also "grandfathering" individuals who have received board-certification prior to NRC's recognition of a specialty board would be in line with the grandfathering for medical use of the new byproduct materials. In certain cases, such as those individuals who have been board certified by the American Board of Health Physics (ABHP) prior to January 1, 2005 and never named as RSO on a NRC or Agreement State license, an individual could not currently be named as an RSO based on their board-certification even though the ABHP made no changes in its certification process to receive NRC-recognition. WU also strongly recommends that NRC allow grandfathering of individuals who were board-certified prior to NRC-recognition for all specialty boards which receive NRC-recognition prior to the required implementation date, August 9, 2009, for the new byproduct definition.

NRC Staff Response: Any revisions to the training and experience requirements would require a revision to NRC's current regulations. Therefore, this comment is beyond the scope of this guidance document revision.

Location	Subject	Comment
General Comment	Distribution of Radionuclides	WU plans to incorporate a commercial radiopharmacy license into our overall broad scope license for the distribution of copper-64 (Cu-64), and possibly other accelerator-produced radionuclides, to other research entities for their production of radiopharmaceuticals for human research use. WU's continued intent in supplying accelerator-produced radionuclides is to further the research and development of imaging techniques with eventual technology transfer to an entity that would commercially produce and distribute one or more of these radionuclides.

Since these research entities, which are located throughout the U.S., do not meet NRC's proposed definition for being in a "consortium" with WU, we will be obligated to become a "commercial" radiopharmacy, even though our distribution of accelerator-produced radionuclides for eventual human use will continue to be for noncommercial research and development. We plan to list separately a license item for Cu-64, and possibly other accelerator-produced radionuclides, and plan to identify purpose of use as 10 CFR 32.72.

Question – In Appendix D.5 (pages D-2 & D-3), the purpose of use is listed as 10 CFR 32.72 and 10 CFR 30.41. NRC has stated in the draft Federal Register Notice (SECY-07-0062, Enclosure 1, p.128):

"In general, a PET radionuclide production facility may transfer excess PET radionuclides to other licensees that are authorized to receive such PET radionuclide transfer under 10 CFR 30.41."

"An applicant's intent regarding noncommercial distribution, transfer, or commercial distribution will be evaluated as part of the licensing review process to ensure that the proper license or authorization is issued."

Does NRC agree a licensee that is required to obtain a commercial radiopharmacy license to cover a subset of its transfer of radionuclides, such as described here for WU's situation, is allowed to make non-commercial transfers under 10 CFR 30.41 for radionuclides not included in commercial radiopharmacy license purpose of use? |

Location	Subject	Comment
		NRC Staff Response: The commenter does not appear to be raising a comment on this guidance document, but rather to be asking a specific question as to licensing its activities. If there are any questions as to what authorizations are needed, licensees should contact their NRC Regional Office.

Location	Subject	Comment
General Comment	Distribution of Radioactive Drugs	What guidance does NRC give license applicants for 10 CFR 32.72 distribution of radionuclides that may contain other radionuclide contaminants? Should not guidance on how to describe these potential contaminants be included in this document? Examples of these types of radiopharmaceuticals that are widely used include: Sm-153 Quadramet which can include Eu-154 and Eu-155 Tl-201 Thallous Chloride which can include Tl-200, Tl-202 and Pb-203 In-111 Indium Chloride which can include In-114m and Zn-65

NRC Staff Response: The NRC understands that some radionuclides may contain small amounts of radionuclide contaminants. Generally, NRC authorizes the possession and/or use of the main radionuclide and assumes that contaminants are part of the main radionuclide's characteristics. Therefore, the NRC staff does not believe that additional guidance is needed for describing radionuclide contaminants.

Location	Subject	Comment
Section 8.5.1 (Page 8-7)	Sealed Sources	The draft guidance discusses what a radiopharmacy applicant should do if it possesses a sealed source containing the new byproduct material and there is no Sealed Source and Device (SSDR) certificate. NRC expects this applicant to provide information required under 10 CFR 30.32(g), which states: "An application for a specific license to use byproduct material in the form of a sealed source or in a device that contains the sealed source must either--
		(1) Identify the source or device by manufacturer and model number as registered with the Commission under § 32.210 of this chapter or with an Agreement State; or
		(2) Contain the information identified in § 32.210(c)."
		10 CFR 30.32(g)(1) appears to be asking for the SSDR, which seems redundant since NRC requests this information because there is no SSDR. To meet 10 CFR 30.32(g)(2), 10 CFR 32.210(c) states:
		"The request for review of a sealed source or a device must include sufficient information about the design, manufacture, prototype testing, quality control program, labeling, proposed uses and leak testing and, for a device, the request must also include sufficient information about installation, service and maintenance, operating and safety instructions, and its potential hazards, to provide reasonable assurance that the radiation safety properties of the source or device are adequate to protect health and minimize danger to life and property."
		The information NRC requests may not be readily available to the applicant if the radiopharmacy purchased the source from someone else. If NRC asks for this information from every applicant possessing the sealed source, then it appears that NRC will be receiving multiple requests to do a safety evaluation for the same sealed source model. We recommend that NRC work directly with the sealed source manufacturers to begin conducting safety evaluations and issuing SSDR certificates. Guidance for applicants who only possess these sealed sources should be able to provide NRC with the manufacturer name, source model number and general physical description.

NRC Staff Response: The information required under 10 CFR 30.32(g)(1) and (2) applies to all sealed sources, devices, and sealed source-device combinations. As part of the NARM rule, a new paragraph (3) was added to 10 CFR 30.32(g) that allows a basis for the licensing of sealed sources and devices containing NARM that were manufactured before the effective date of the rule and for which all of the information required in 10 CFR 30.32(g)(1) and (2) is not available. Without this provision, an applicant who wanted to use the NARM source or device that was not registered in the SSDR would have been required to submit all of the safety information identified in 10 CFR 32.210(c), because this information had not been submitted already by the manufacturer or distributor as part of registering the source or device. When all the information required by 10 CFR 32.210(c) is not available, 10 CFR 30.32(g)(3) allows a basis for licensing these sources and delineates information that will be required to license a NARM source or device. The NRC recognizes that a number of "legacy" sources containing these materials were produced by manufacturers that are no longer in business or have stopped making the sources and/or devices some time ago. These are the sources for which NRC expects to receive information under the provisions of 10 CFR 30.32(g)(3). The text in this section of the guidance document has been revised to clarify this new provision.

Location	Subject	Comment
Section 8.6.1 (Page 8-12)	Verification of License Authorization	The draft document provides guidance on verifying whether a transferee is allowed to receive the type, form and quantity of byproduct material to be transferred. Supplying copies of licenses has become problematic in the security conscious world of NRC Increased Controls. In NRC's RIS 2005-31, "Control of Security-Related Sensitive Unclassified Non-Safeguards Information Handled by Individuals, Firms, and Entities Subject to NRC Regulation of the Use of Source, Byproduct, and Special Nuclear Material," Appendix 3, material licensees are told to withhold authorized quantities, manufacturers, model numbers and locations of sealed sources and devices exceeding threshold values. For some licensees, like WU, supplying a copy of the NRC license with multiple areas blacked out can look unprofessional and suspicious. **Comment & Recommendations** – If NRC states the radiopharmacy should "verify that the address to which radioactive materials are delivered is an authorized location of use listed on the customer's license," and notes that the "most common form of verification" is possession of a "valid copy of the customer's NRC or Agreement State license", we are concerned that licensees will only accept copies of licenses as verification. We **recommend** either NRC delete mention of obtaining a copy of the license, or expand the explanation that another acceptable verification is a written certification by the licensee receiving the radioactive material that states the licensee is authorized by license or registration to receive the type, form, and quantity of byproduct material to be transferred, specifying the license or registration certificate number, issuing agency and expiration date. We also **recommend** that NRC include in this discussion that some licensees may choose to provide their own written verification and not to provide a copy of their license based on NRC guidance given in RIS 2005-31.

NRC Staff Response: This comment is not related to the NARM rule and, therefore, is beyond the scope of this guidance document revision. This comment will be evaluated during any future revision of this NUREG..

Location	Subject	Comment
Chapter 3 (Page 3-1)	Management Responsibility	Definition of "Management" should be similar to that found in Vol. 11 (Broad Scope). We suggest it be modified to read: " 'Management' refers to the processes for conduct and control of a Radiation Safety Program and to the individuals who are responsible for those processes and have authority to provide necessary resources to ensure safety and to achieve regulatory compliance."

NRC Staff Response: Changing the definition of "Management" for this guidance document is beyond the scope of this revision. The definition for "Management" found on page 3-1 of this guidance document is consistent with other NUREG-1556 guidance documents (e.g., NUREG-1556, Volume 12).

Location	Subject	Comment
Section 8.5.1 (Page 8-8)	Unsealed Byproduct Material	To strengthen the idea that this draft document has been updated to include the new byproduct materials, we suggest that iodine-123 be included as an example for potentially volatile materials.

NRC Staff Response: Iodine-123 has been added as an example for potentially volatile materials in this section.

Location	Subject	Comment
Section 8.7.2 (Page 8-20)	Authorized Nuclear Pharmacist	Should the statement, "For an individual qualifying under 32.72(b)(5)" be corrected to reference 32.72(b)(4)?

NRC Staff Response: Changes have been made to the document to implement this correction.

Location	Subject	Comment
Section 8.9.1 (Page 8-29)	Facilities and Equipment	The two bulleted items following Figure 8.3 should be deleted since they are repeated text.

NRC Staff Response: The two bulleted items following Figure 8.3 have been removed.

Location	Subject	Comment
Section 8.10.6 and Appendix C (Pages 8-45, 8-47, and C-11)	Safe Use of Radionuclides	To strengthen the idea that this draft document has been updated to include the new byproduct materials, we suggest that performing Sr-82 and Sr-85 breakthrough measurements also be included for elution from a Rb-82 generator.

NRC Staff Response: The NRC only has specific breakthrough test requirements for molybdenum-99/technetium-99m and strontium-82/rubidium-82 generator systems under 10 CFR 30.34(g). However, the strontium-82/rubidium-82 generator breakthrough test is not generally performed at the pharmacy, but at the medical facility prior to first patient use. Therefore, this guidance document only refers to the molybdenum-99 breakthrough measurements.

Location	Subject	Comment
Appendix G (Page G-4)	Typographical Error	The "[BOLD]" after IV. Recentness of Training should be deleted.

NRC Staff Response: The term "[BOLD]" has been deleted from this section.

Location	Subject	Comment
Appendix G (Page G-5)	Training and Experience Documentation	The Note on this page states, "An individual that is board eligible will not be considered for this pathway until the individual is actually board-certified." Does NRC consider an individual to be board-certified when they have received written confirmation that they successfully completed their board exam?

NRC Staff Response: Individuals are considered board-certified when they receive written confirmation from the specialty board that they are certified. Successful completion of the board exam may not mean that the individual is board-certified.

NRC FORM 335 (9-2004) NRCMD 3.7	U.S. NUCLEAR REGULATORY COMMISSION	1. REPORT NUMBER (Assigned by NRC, Add Vol., Supp., Rev., and Addendum Numbers, If any.)
BIBLIOGRAPHIC DATA SHEET *(See instructions on the reverse)*		NUREG-1556, Volume 13, Rev. 1

2. TITLE AND SUBTITLE		3. DATE REPORT PUBLISHED	
NUREG-1556, Volume 13, Rev. 1, Consolidated Guidance About Materials Licenses: "Program-Specific Guidance About Commercial Radiopharmacy Licenses"		MONTH	YEAR
		November	2007
Final Report		4. FIN OR GRANT NUMBER	
		NA	

5. AUTHOR(S)	6. TYPE OF REPORT
Duane E. White, NRC/FSME Janine F. Katanic, NRC/Region IV Donna-Beth Howe, NRC/FSME	Final
	7. PERIOD COVERED *(Inclusive Dates)*

8. PERFORMING ORGANIZATION - NAME AND ADDRESS *(If NRC, provide Division, Office or Region, U.S. Nuclear Regulatory Commission, and mailing address; if contractor, provide name and mailing address.)*

Office of Federal and State Materials and Environmental Management Programs
U.S. Nuclear Regulatory Commission
Washington D.C. 20555

9. SPONSORING ORGANIZATION - NAME AND ADDRESS *(If NRC, type "Same as above"; if contractor, provide NRC Division, Office or Region, U.S. Nuclear Regulatory Commission, and mailing address.)*

Same as above

10. SUPPLEMENTARY NOTES

11. ABSTRACT *(200 words or less)*

On August 8, 2005, the Energy Policy Act of 2005 (EPAct) gave NRC new regulatory authority over additional byproduct material. This new byproduct material now also includes naturally occurring materials, such as discrete sources of Radium-226, and accelerator-produced radioactive materials (NARM). This revision of NUREG-1556, Vol. 13 adds guidance needed to license commercial radiopharmacies as a result of the regulatory changes made by the EPAct and the NARM rule, "Requirements for Expanded Definition of Byproduct Material."

This guidance document contains information that is intended to assist applicants for commercial radiopharmacy licenses in preparing their license applications. In particular, it describes the type of information needed to complete NRC Form 313, "Application for Materials License." This document both describes the methods acceptable to NRC license reviewers in implementing the regulations and the techniques used by the reviewers in evaluating the application to determine if the proposed activities are acceptable for licensing purposes.

12. KEY WORDS/DESCRIPTORS *(List words or phrases that will assist researchers in locating the report.)*	13. AVAILABILITY STATEMENT
Accelerator Commercial radiopharmacy PET NARM Guidance	unlimited
	14. SECURITY CLASSIFICATION
	(This Page)
	unclassified
	(This Report)
	unclassified
	15. NUMBER OF PAGES
	16. PRICE

NRC FORM 335 (9-2004)

PRINTED ON RECYCLED PAPER

www.ingramcontent.com/pod-product-compliance
Lightning Source LLC
Chambersburg PA
CBHW081432170526

45166CB00008B/2180